D0924290

PRINTING PRESSES

PRINTING PRESSES

HISTORY AND DEVELOPMENT FROM THE FIFTEENTH CENTURY TO MODERN TIMES

JAMES MORAN

UNIVERSITY OF CALIFORNIA PRESS
BERKELEY AND LOS ANGELES

University of California Press
Berkeley and Los Angeles

First Paperback Edition 1978
ISBN: 0-520-02245-9 clothbound
0-520-02904-6 paperbound
Library of Congress Catalog Card Number 72-75519

Printed in the United States of America

1 2 3 4 5 6 7 8 9 0

Front endsheet: The workshops of Hughes and Kimber, printing machinery manufacturer, from an undated catalog (c. 1880).

Back endsheet: An illustration from a 19th century encyclopedia showing the various types of printing machines available.

CONTENTS

ILLUSTRATIONS

FIGURES IN TEXT

A number of these illustrations began life as publicity material and have been repeatedly reproduced during the 19th and 20th centuries in books, brochures, catalogues, trade-paper articles and advertisements. Other sources include block and advertisement proofs. Some are from the author's own collection but additional acknowledgment is made to: Lee Augustine, Fernand Baudin, Alan Dietch, Otto Lilien, Vivian Ridler, Harry Rochat, Rollo Silver, Martin Speckter, Dr. Berthold Wolpe; Faber & Faber Ltd.; the St. Bride Institute Printing Library and the Printing Historical Society.

THE PLATES

following page 224

INTRODUCTION AND ACKNOWLEDGEMENT

THIS BOOK DEALS with the development of the relief printing press from its inception in the middle of the fifteenth century until approximately 1940, when it may be said to have reached its zenith. I occasionally mention developments after that date, but the five hundred years happily make a neat and convenient period to cover. During the world war of 1939–45 and for some ten years thereafter there was little new in printing-press manufacturing technique, but during the next decade the seeds were sown for completely new approaches which will, no doubt, come to fruition in the last quarter of this century. I have resisted the temptation to prophesy at any length about these changes, firstly because it is impossible to be specific, and secondly, if previous experience is anything to go by, many of the existing types of press will continue in operation for many years to come.

I have also limited myself to the relief or letterpress press and machine, and have not dealt with either the lithographic or gravure processes. Much of the development in these two processes as far as the press was concerned was on the same lines as that in relief printing and (particularly with regard to lithography) the greatest changes may yet be to come.

If, therefore, another edition of this work is ever called for it might provide the opportunity for bringing it up to date at that point of time with developments not only in the three well-known printing processes, but also in the new machines for mass reproduction, which, however, may not really justify the description of 'presses'.

While there have been books published on this subject over a period of nearly three hundred years I think I may claim that this is the most comprehensive, but, even so, in order to keep the book within reasonable bounds, I have had to omit descriptions of some presses, which were made in large numbers in the late nineteenth and early twentieth centuries. The book may also not be free from error, and I would therefore be grateful if readers would point out any mistakes and would also let me have details of any press or presses which represented a major step forward and which I may have failed to mention.

In this work I have been considerably assisted by other enthusiasts in various parts of the world and I would like to acknowledge my indebtedness to them. I have encountered nothing but the most pleasant co-operation and I would be very remiss if I did not state that quite a number of them have shared the results of their researches with me. Libraries and historical associations, manufacturers and trade bodies have been most helpful. I would like to record my thanks to the following: Virginia M. Adams, Providence Public Library, Rhode Island; Maurice Audin, Musée de l'Imprimerie et de la Banque, Lyons; Lee Augustine of Cincinnati, a major source of information and inspiration for many years; Christian Axel-Nilsson, Nordiske Museet, Stockholm; Ronald Ayers, Adana (Printing Machines) Ltd., for a great deal of help; Fernand Baudin, for assistance on Belgian sources; Dame Beryl Paston Brown, Principal, Homerton College, Cambridge; David Chambers; W. E. Church, Science Museum, London; J. D. Cleaver, Oregon Historical Society; Muir Dawson, Los Angeles; Alan Dietch, Rochester, N.Y. for a wide range of assistance; A. G. Davies, Hertford Museum; Dr. James Eckman, Rochester, Minn.; Kenneth Hardacre; Elizabeth Harris, Smithsonian Institution, Washington, D.C., for valuable assistance on American inventions; Peter Henson, the Monotype Corporation; Henry Beetle Hough, Edgartown, Mass.; G. Allen Hutt, for both information on Russia and on newspaper press developments; André Jammes, Paris; J. P. J. R. Kidder, the Uckfield Press, Sussex; Roger Levenson, Berkeley, California, particularly for research on Columbian presses; E. G. Lindner, an indefatigable searcher after early presses; Otto Lilien; J. & J. S. Mackay, Morpeth; Miehle-Goss-Dexter Ltd., London; Professor Jack Morpurgo; Eric Mortimer, Singapore; C. E. Mosher, the Kelsey Co., Meriden, Conn.; S. Ó. Conchubhair (now Librarian, Kildare County Library); Leslie Owens, principal of the London College of Printing; Dr. Willem Ovink, Amsterdam; Stephen R. Parks, Curator, Osborn Collection, Yale University Library; George Painter, British Museum; Robert C. Pettit, Nebraska State Historical Society; Lewis J. Picton, Cape Town; Roy Pilkington, Boston, Lancs.; William C. Pollard, Librarian, College of William and Mary in Virginia, for information from the Ralph Green Collection in the Earl Gregg Swem Library; Andrew A. Polscher, Detroit, for help on early American presses; Vivian Ridler, Printer to the University of Oxford, for generously lending me the results of his own researches; B. W. Robinson, Victoria and Albert Museum; the always helpful staff of the St. Bride Institute Printing Library; Rollo Silver, distinguished printing historian for never hesitating to answer questions and for passing on information from his own studies; George Simpson, of Swindon, Wilts.; Martin Speckter, of New York, a major source of information on miniature presses; Reynolds Stone; Michael Turner, of the Bodleian Library; the Vandercook Division, Illinois Tool Works Inc., Chicago; P. J. van der Walt, City Librarian, Kimberley; S. F. Watson; Roby Wentz, of California, source of information about early Western presses; Edwin Wolf II, Philadelphia; John Mason; Godric E. S. Bader, for information concerning his great-grandfather, Henry Ingle; Vance Gerry, The Weather Bird Press, California; Harry F. Rochat, printers' engineer; Ian Stephens, Northampton; Miss P. Adams, New Zealand Historic Places Trust, Wellington, N.Z.; Noel Edwards, Russell, N.Z.; A. G. Bagnall, Chief librarian, The Alexander Turnbull Library, Wellington, N.Z.; and to Dr. Berthold Wolpe, inspirer, guide and friend.

JAMES MORAN

Engraved for the New Universal Magazine, 1752.

B. Cole sculp.

1. The printing press in this eighteenth-century representation of a printing house is basically the structure and mechanism used by printers for four hundred years. The various working parts are described in the following chapter, but this press carries what became known as the Blaeu hose (see pages 31–2). Despite the assurance that it is a 'true representation' the illustration is not completely accurate, B. Cole, the engraver, sharing with other artists over the centuries, including the great Albrecht Dürer, an inability to render details correctly. No depth is given to the forme, which the 'beater' (G) is inking; and the press has been depicted the wrong way round. The rounce for winding the forme under the platen, the ink block and pegs for the ink balls should be on the pressman's side and the bar should curve in the opposite direction. The tympan (H) has no frisket attached, and, in any case, there is no provision for a stay to hold the frisket when thrown back after a printing operation

1

THE BEGINNING

THE MAKING OF an imprint—transferring a mark by pressure—is as old as mankind. Printing, in the sense of impressing or stamping one surface with another, which has been incised or cut in relief and then inked, goes back at least to A.D. 175, when it was practised by the Chinese, who had invented paper some seventy-five years earlier.

In the most familiar method, blocks of wood were cut in relief and inked with a water-based ink. Paper was laid on the block and gently rubbed with a bamboo stick, bone or dry brush to produce an impression. This art spread to other parts of the Orient, eighth-century Japanese and Korean printed texts still surviving. By 1041 the Chinese had produced individual relief ideographs made of earthenware; by 1300 the Uigur Turks, on the borders of Turkestan, had cut similar individual characters in wood; and the Koreans, from 1300, were casting individual relief characters in bronze, probably in sand from wooden patterns.

Why, therefore, is printing considered to be a mid-fifteenth-century European invention? The method of production of the fifteenth-century European block book was basically not much different from that which had been carried out in the Orient for some hundreds of years. Typographic printing, on the other hand, with its special characteristics, can be truly said to be a European invention. The Koreans came nearest to developing this form of printing, but did not progress beyond primitive sand-casting of their characters, which, in any case, did not lend themselves to mass-reproduction techniques. For it was the lack of an alphabet of a limited number of characters which made this mass-production of types impracticable. It does not require much effort of the imagination to realize the difficulties involved in casting the thousands of ideographs required in Chinese writing; nor, at that period, was there much incentive to overcome the difficulties. Such official communications as were thought necessary could be written by hand; printing had only been considered, by the Chinese, at least, as a method of recording decrees permanently, and had little to do with public demands for reading matter.

Europe, fortunately, had its twenty-four-letter alphabet, which derived from the Romans, Greeks and Phoenicians, and it only needed a satisfactory method of casting the letters in relief for the way to be open for the reproduction of identical texts in quantity. An effort was made in Holland, possibly in the first part of the fifteenth century, to cast types in sand, but the results were crude and the method so time-consuming that it could not have led to modern printing from type, which is consequently attributed to Johann Gutenberg of Mainz (*c.* 1398–1468) who began experimenting some time before 1440.

The second aspect of the invention, following the existence of an alphabet, concerned the way in which Gutenberg manufactured type. He sought a more efficient method than sand casting, which is appropriate to large objects but not to delicate alphabetic characters. He was connected with the Mainz mint, in a hereditary capacity, and presumably would have been familiar with the methods of cutting a punch, sinking a matrix, and casting metal in a mould. His key invention consisted of an ingenious combination of mould and matrix. A letter was engraved in relief on a hard-metal punch, which was struck into a small slab of softer metal to provide a matrix—the intaglio version of the letter. The matrix was placed at the bottom of a mould capable of casting shanks of metal of the same height, but which could be adjusted horizontally to accommodate individual matrices of varying widths, that is letters from the narrowest to the widest. Molten metal was poured into the mould, and allowed to cool, the resulting product being a character in relief at one end of a shank of metal, long enough to be grasped between finger and thumb. Such types could be composed into words and lines, and being of constant height could be locked into a frame (the chase) to present a uniform and rigid printing surface, known as a forme.

The third aspect concerned the actual method of printing, involving something new—a printing press, which was a fundamental necessity if the new process was to succeed. Gutenberg needed an ink which would not reticulate on his metal type, and his solution was to adapt the discoveries of painters of some decades earlier by using an ink made from pigment ground in a linseed-oil varnish. Rubbing back and forth over such an ink in the Chinese style would have produced a smeared impression. He also needed paper, which had been manufactured in Europe for some years, but it was a paper made from macerated linen and cotton cloth, treated with size to give it a hard, opaque surface for writing on with a quill pen. It was unlike the thin, soft, pliable and absorbent paper used with water-based ink in Oriental rub-printing. The constitution of both ink and paper therefore required a printing method whereby a sharp but decisive impact could be made between paper and type. The well-sized paper posed a particular problem, which was overcome by damping it before printing. Apart from the technical reasons, the very nature of Gutenberg's typographical invention demanded a new printing device. He was a pioneer of mass-production—the manufacture of exactly similar relief letters on metal stems, and he had to find a more rapid method of manufacturing printed texts than the slow rub-printing if his invention was not to be nullified.

Gutenberg was, therefore, forced to become a pioneer in tackling the problem of keeping one printing sub-process in step with another. His solution was to last some four and a half centuries, when the problem then involved the increase in speed of composition of type to match the demands of the rotary printing press; and it is likely that in the next few decades the position will once again be reversed, as computerized composition overtakes traditional printing methods.

What Gutenberg needed was a mechanical device which would perform for a large area of type the function of the human arm when it pressed a small seal on to paper. In this case the Mint, which had provided the inspiration for casting type, was of no help, for in Gutenberg's day the human arm still provided much of the motive power. Smaller coins were made by hammermen, who struck dies into metal blanks. The coin press was centuries away, providing an inspiration for Wilhelm Haas, who devised a new press in 1772. A press of some sort was obviously what Gutenberg needed. At his time there were two main kinds, both of ancient origin, in use for a variety of domestic, agricultural and craft purposes. The first was based on the lever and fulcrum and the other on the screw. The beam press, developed before 1500 B.C., utilizes a lever, of the second order, one end of which is captive in a recess and the other free to be drawn down by manpower or weights to press anything placed between it and a platform. By the first century A.D. a screw had been introduced to replace the weights, to screw down the free end of the lever. Beam presses were particularly suited to large-scale operations—for the pressing of olives and grapes, while for pressing smaller quantities—oil seeds, herbs or sheets of papyrus—the lever was eliminated, so that the screw bore directly down from above, in the manner of the modern copying press. This was in use in Europe from at least the first century A.D. Either mechanism might have been chosen as the basis of the printing press, but it is apparent that Gutenberg rejected the beam press in favour of the downward-acting screw.

The screw is, in effect, an inclined plane carried round a column, but in the application of the screw the weight of resistance is not, as in the inclined plane and wedge, placed on the surface of the plane. The power is usually transmitted by causing the screw to move in a hollow cylinder, on the interior surface of which a spiral cavity is cut, corresponding exactly to the thread of the screw, and in which the thread will move by turning the screw round continually in the same direction. The hollow cylinder is usually called the nut, and one way of urging the screw round is by a lever or bar screwed into the nut. When the nut is turned by means of this bar it advances in the direction of its length and urges anything attached to it downwards, pressing any substance placed between it and a fixed platform below. This, in essence, was the kind of screw mechanism adopted for the printing press.

It has been a tradition, particularly in Germany, that Gutenberg adapted a wine press, and in 1840 for a book produced in Paris to celebrate the quatercentenary of the invention of printing, entitled *Histoire de l'invention de l'Imprimerie par les Monuments*, Eugène Duverger, the publisher, invented a series of letters from Gutenberg to a Franciscan friar, Frère André, in which an attempt was made to transform tradition into fact. Gutenberg, frustrated in finding a method to replace rub-printing, is made to describe a flash of inspiration when attending a wine harvest. He studied the power of the wine press and it suddenly occurred to him that the same pressure might be applied to type to impress an image on paper.

Duverger only hints that the letters are not genuine, but his friend Charles Nodier, in a complimentary review in *Bulletin du Bibliophile*, August–September 1840, really gives the game away, doubting whether the letters ever existed. They have been accepted as genuine by Arthur Koestler, in his book *The Act of Creation* (1964), and in a subsequent television programme. Koestler, who calls the correspondent 'Frère Cordelier', probably accepted the letters as they back up his theory of 'bisociation' of ideas.

However, while there is nothing inherently improbable in the idea that the frame and screw of the wine press was used as the basis of the first printing press (even if Frère André's letters are the product of a nineteenth-century publisher's imagination) Gutenberg may also have been acquainted with the papermaker's press, although it is likely that he bought his paper from a merchant and may never have entered a paper-mill. But casting round for something which provided downward pressure, Gutenberg could have seen a screw press in his own home—a linen press, for example. There were also others to press cloth. Much really depends on Gutenberg's acquaintance with various trades and crafts. It may be that he needed no prompting, and simply took for granted that a screw press, a familiar object, could serve his purpose, utilizing the kind which was motivated by the pull of a bar rather than one which worked by twisting a horizontal rod, as being more convenient. At the same time it is misleading to state that Gutenberg merely 'adapted' a screw press, inferring that this was a simple feat. In fact, there were considerable differences between screw presses and the printing press which finally emerged.

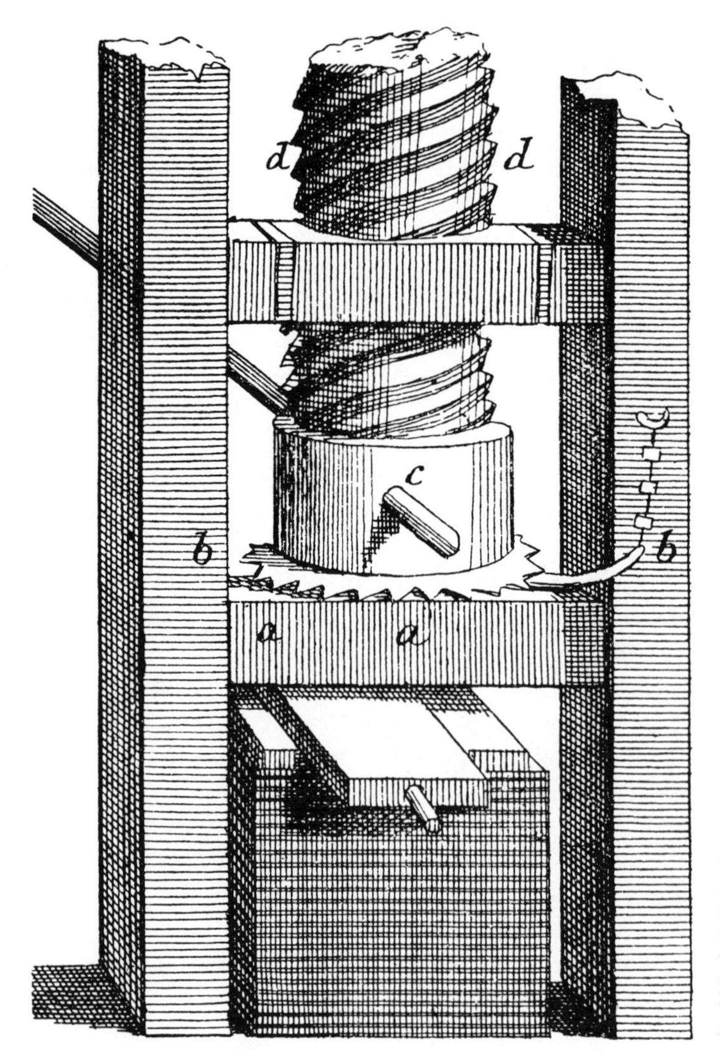

2. The papermaker's press remained unaltered for centuries. This 1762 engraving shows the heavy wooden screw (d), the holed collar (c) and the pressing board (a). To prevent the screw from turning back a ratchet wheel and ratchet (b) was provided

A particular point to bear in mind is that the massive wine and paper presses were capable of great pressure, and, as such, could not possibly have been suitable to print the books for which Gutenberg and his associates, Johann Fust and Peter Schoeffer, are held responsible. These books were intended to rival the best hand-produced manuscripts of the day, and a much more sensitive instrument was needed to print, for example, the magnificent pages of the 42-line Bible and the delicate sectional initials of the Mainz *Psalter*.

In the so-called *Cologne Chronicle*, published in 1499, Ulrich Zell, an early source of information, reports that Gutenberg's discoveries took place in 1440, and this is the conventional date given to the invention of typographical printing; but from documents relating to a Strasburg lawsuit, in which Gutenberg was involved, it can be deduced that he was developing his invention perhaps two or three years earlier. In the documents, a press is mentioned several times, although there are no details of its construction. Konrad Saspach, a wood turner, gave evidence that he began building the press after the death in December 1438 of Andreas Dritzehen, one of Gutenberg's partners, and that it took him three weeks to complete the work. Gutenberg, it appears from the documents, was at pains to make sure the press was shown to nobody outside the partnership, which is a slight indication that it could not have been an ordinary domestic press.

The form of the first printing press, which did not change in essence for nearly four hundred years, must have resulted from a series of experiments initially by Gutenberg and Saspach, and then by printers in the next few decades. Basically the press used in vineyard, paper-mill and bindery consisted of two main uprights with cross-pieces at top and foot. Through the top cross-piece there penetrated a turned wooden screw which was encircled by a nut in the form of a collar with a series of holes in its circumference. The screw terminated in a flat pressing board which slid between the uprights. Pressure was provided by pulling a pole inserted in one of the holes in the collar, causing the screw to move downwards; increased pressure was obtained by moving the pole from one hole to another. The need for modification of such a press for the printing process would have soon become apparent. Since continuous squeezing pressure was not required but rather a forceful but brief 'dwell' of the flat board (or platen, as it became known) on the type, the mechanism of the detachable pole and holed collar could be dispensed with, being replaced by a fixed bar to be pulled once only to activate the screw. The comparatively short distance the platen was required to travel would have suggested a shorter screw with an increased pitch.

It is certainly possible to print with a press consisting of a screw with a platen attached—such as a linen or office copying press—but it is a laborious process. Moreover, the pressure exerted by a heavy board by screwing it down may damage the type, and so the need for more resilience produced the idea of freeing the platen from the screw and framework. At some point the platen was freed and hung on hooks at the end of a box-like contrivance (the hose). The screw itself was refined into a spindle with a thread at the top, becoming smooth lower down and gradually tapering into a rounded point (the toe). The spindle turned independently within the hose (avoiding twisting the platen and slurring the impression), the toe thrusting into an indentation in the top of the platen (the stud).

3. First known illustration of a printing press, 1499. (See page 25)

When one considers the clean, even presswork of the early printed books it is inconceivable that the inked forme with a sheet of paper laid on top could have been manœuvred under the platen without slurring or disturbing the type. Some kind of platform or table on which the forme could be positioned accurately and then slid under the platen would have suggested itself. This may have taken time to develop to its final form. At first it is thought that there was simply a flat plank protruding from beneath the platen and held up at its remote end by a forestay. The forme was placed in a box and, after the type had been inked, the sheet of paper may simply have been dropped on to it, and then the box slid along the plank to its position under the platen. This way of placing the paper would have been a delicate proceeding if the page was to be exactly square on the paper or, to use the printers' term, 'in register'. 'Register' also refers to the backing up exactly of type on both sides of a sheet, and to the precise printing of more than one colour without overlapping, so that a more efficient method of guiding the forme was needed than that of placing the paper by simple judgement of hand and eye. In time, therefore, a more elaborate device for bringing the type and paper under the platen evolved. The plank remained, but the box, which acquired the name of 'coffin', was fixed to it. In the coffin a stone was embedded in bran, or similar material, to provide a surface for the forme to rest on, with angle plates at the corners to keep it in position. On the underside of the plank there were fitted cramp-irons, which could slide along the ribs of a frame, known as the carriage. This movement was usually accomplished by a spit mechanism, the turning handle of which was known as a rounce.

A possible confirmation that there was not always a mechanism to bring the forme under the platen in early presses is provided by an illustration of a printing office in a seventeenth-century translation, by Charles Hoole (1659), of a Latin teaching book *J. A. Commenii Orbis sensualium pictus,* which sought to teach Latin by means of pictures depicting everyday trades. Instead of a movable carriage and rounce mechanism on the press there is a long bench, on top of which is a hinged box, which was presumably pushed under the platen.

To obtain evenness of pressure and to protect the type it would also have been desirable to have an interlay of soft material between paper and platen. Thus was developed the tympan, which eventually consisted of an inner and an outer frame, over each of which linen or vellum was stretched, one fitting inside the other. Sheets of cloth ('blankets') or similar material, were laid between the parts, which were then clasped together, the whole tympan being hinged to the coffin. The sheet of paper was positioned precisely on the tympan by pressing it on to points which perforated the margins, the holes serving as guides to exact register when either printing on the other side of the sheet ('backing up') or when printing with an additional colour.

The cleanliness of early printing also indicates some form of protection sheet—the forerunner of the 'frisket'. This was a piece of parchment, or paper, with windows cut in it to allow the type to print through while, at the same time, protecting the paper from ink which may have adhered to the other parts of the

4. Device of Jocodus Badius Ascensius (1507). The screw is working in a hose and the platen is free from the sides of the press, but the method of attachment to the hose is not clearly depicted. The spit mechanism for bringing forme under platen is shown. (See page 25)

5. Device of Jocodus Badius Ascensius (1520) shows the influence of the wood carver (see South German press, page 35). The platen can be seen hanging from the four corners of the hose. Cap and screws have been dispensed with and, instead, adjustable jacks extend from head to ceiling. There was still a need for a brace (top right) to steady the press. Tools are attached to the head—dividers to measure page margins, bodkin for pricking point holes, scissors and paste brush for cutting frisket and pasting it to the frisket frame. (See 12, page 36)

forme. It was stretched over a thin metal frame which was hinged to the tympan, being folded over the sheet of paper before printing. The frisket thus also assisted in holding the paper in position. The tympan and frisket may well have been a very early development as pages of a book of 1487 show marks or 'bites' from a frisket frame.

The earliest mention of the terms 'tympan' and 'frisket' is probably that of Plantin in 1567, by which time they must have been a well-established part of the printers' vocabulary.

It is not known exactly when each of these small, but important, technical innovations took place. Each may have developed of its own accord in response to given needs, until it was realized that they could be combined into what is known nowadays as a 'multi-purpose' unit, catering for 'make-ready', correct register and clean printing.

Those who reconstructed the wooden press in the Gutenberg Museum at Mainz incorporated these refinements, working on the hypothesis that the quality of Gutenberg's work demanded them, although details of his press are not known. We rely for evidence as to what the earliest printing presses were like on the work of artists, but since these men were often not strictly technical illustrators, their representations are sometimes not reliable.

The first known illustration of a printing press was certainly not drawn to enlighten future generations as to its characteristics. It appears in an edition of the *Danse Macabre*, published in Lyons by Mathias Huss in 1499. Death is depicted carrying off a printer and a bookseller, and, such as it is, we may take it that the cut illustrates a French fifteenth-century printing office. Unfortunately, although the general construction of press can be made out, the very aspect which would have been of most interest—the way in which the platen was hung—is obscured by the struggling figure of the pressman. However, the illustration does show clearly the supports, or stays, between the top of the press and the ceiling, which were found to be necessary to keep the press stable; a coarse wooden screw, and a straight pole or bar. Particularly interesting is the plank held up by a stay and on which there is a box, to which we may presume a tympan is hinged by what look like leather straps. No winding mechanism is visible and it may be conjectured that the box was pushed under the platen by hand at this date. The other pressman (or 'beater') is holding an ink-ball, which hardly changed in appearance until it was replaced by a roller some three hundred and fifty years later. Two ink-balls were used to ink the forme. They were made of untanned leather or sheepskin, stuffed with wool or hair, and nailed around a wooden handle or stock. Ink was spread out on to a slab and rubbed out thinly with a wooden device known as a brayer.

The little rest, or gallows, gives additional credence to the idea that there was a tympan to be thrown back on it when the forme was being inked. The unusual position of the pressman, who usually stood next to his companion, is probably the result of the artist's license as he wanted to show the figure of Death full face.

About forty representations of the printing press before 1600 are known, some of them, however, being adaptations of others. They vary in value as evidence as to the nature of the early press, but there are a few which, taken together, indicate that the basic elements were established by the beginning of the sixteenth century. The next earliest known illustration (1507) is that of Jocodus Badius Ascensius, of Paris, who shows the printing press in a series of devices, all based on one design, which was copied by other printers. In the 1507 device the rendering of the platen is crude but the presence of the hose indicates that by this time it must have been hung rather than attached to the screw. The transverse beam (the head), which is penetrated by the screw, is now connected with an additional cross-piece (the cap) by two long bolts. This is the first indication that in some presses the head could be moved up and down to suit the required pull. According to Joseph Moxon, in his *Mechanick Exercises* (1683–4), when the press was assembled the joints of the head were soaped and greased to aid movement. Depending on the strength of the pressman and the size of the work to be printed the pull could be made longer or shorter. The screws were undone and the head moved downwards so that packing could be put into the mortises—blocks of wood for a hard pull, felt or card for a soft pull—and then screwed up tightly again.

The pressman can be seen turning a handle, and in the 1512 device of Dirk van den Barner, of Deventer, a similar handle can just be seen. The primitive method of pushing the forme under the platen had clearly, by the early sixteenth century, given way to a movable bed, operated by handle, drum and straps. This mechanism is not always clearly indicated in early illustrations. It was, in effect, an iron spit extending across the width of the press beneath the carriage. In the middle was a drum or barrel upon which were wound two girts or straps, which ran in either direction the length of the carriage. One end of each was fixed to the barrel and the other to either end of the carriage. Winding the handle, or 'rounce', would bring the type forme under the platen and in the other direction bring it out again.

Badius's device is additionally revealing. Here the platen is obviously hanging, and the toe of the spindle is pushing into the top of the platen. The presence of scissors and brushes may indicate that a frisket for protection was cut, but it is not until the illustration in Stumpf's *Schweizer Chronik*, published in 1548, that we see a frisket frame hinged to the tympan. In this illustration also seen faintly are the press points on to which the sheet was thrust. Such points marked the margin of the 42-line Bible and must have been an early development. They become much clearer in Amman's copy of this cut in *Stände und Handwerker*, Frankfurt, 1568. For the first time, also, in Amman's version we see a frisket-stay—a rope attached to the ceiling, which was later sometimes replaced by a wooden batten.

Metal began to be substituted for wood in making certain parts of the press, but when this process began is not certain. Not a great deal of time could have elapsed, however, as a letter written in 1481 by the printer Adolf Rusch to Johann Amerbach, of Basle, asked for four platens to be cast for him. In Plantin's *Dialogues* the platen is called a large piece of iron. Leonhard Danner, of Nuremberg, is said to have originated the metal spindle in 1550, but in the illustration in the *Schweizer Chronik* of 1548 the tip of the screw protruding from the head certainly looks as if it is meant to be of metal, as does the platen, and possibly the bar with its bulbous end. This bar is slightly curved, but a better example occurs in a publication of the same year, *The Ordinary of Christians*, printed by Anthony Soloker, in London. The thin curved metal bar would have provided a certain springiness, easing the jerk on the pressman's arm. The wooden sleeve was provided to give a better grip.

News of small technical improvements no doubt took some time to spread, and the rate of their adoption was not uniform. Printers in the Low Countries seemed to be ahead of others in the use of metal platens and even beds. While in *Nuovo teatro di machine* (Padua, 1607) Vittorio Zonca recommends a spindle and platen of brass and his illustration indicates a metal spindle, the winding apparatus looks rather crude, the girts being made of rope rather than of leather. There appears to be no stay to hold up the end of the plank, a small table occupying its place. Since it would be dangerous to have no stay at all, it looks as if the end of the plank is held up by a batten fixed to the wall.

An illustration which may show the extent to which improvements were delayed, or on the other hand may simply be the result of an artist's error, appears in a book of engravings, entitled *Nova Reperta*, published in Antwerp in about 1600. The engraving in question shows an Italian printing office. The engraver, Johannes Stradanus, although born in Bruges in 1523 spent almost all his adult life in Italy. The press shown is archaic for 1600. The screw has a holed collar and removable bar, and the heavy platen is fixed to the end of the screw. Particularly noticeable is the lack of winding mechanism. In its place is a knob and ring, presumably to

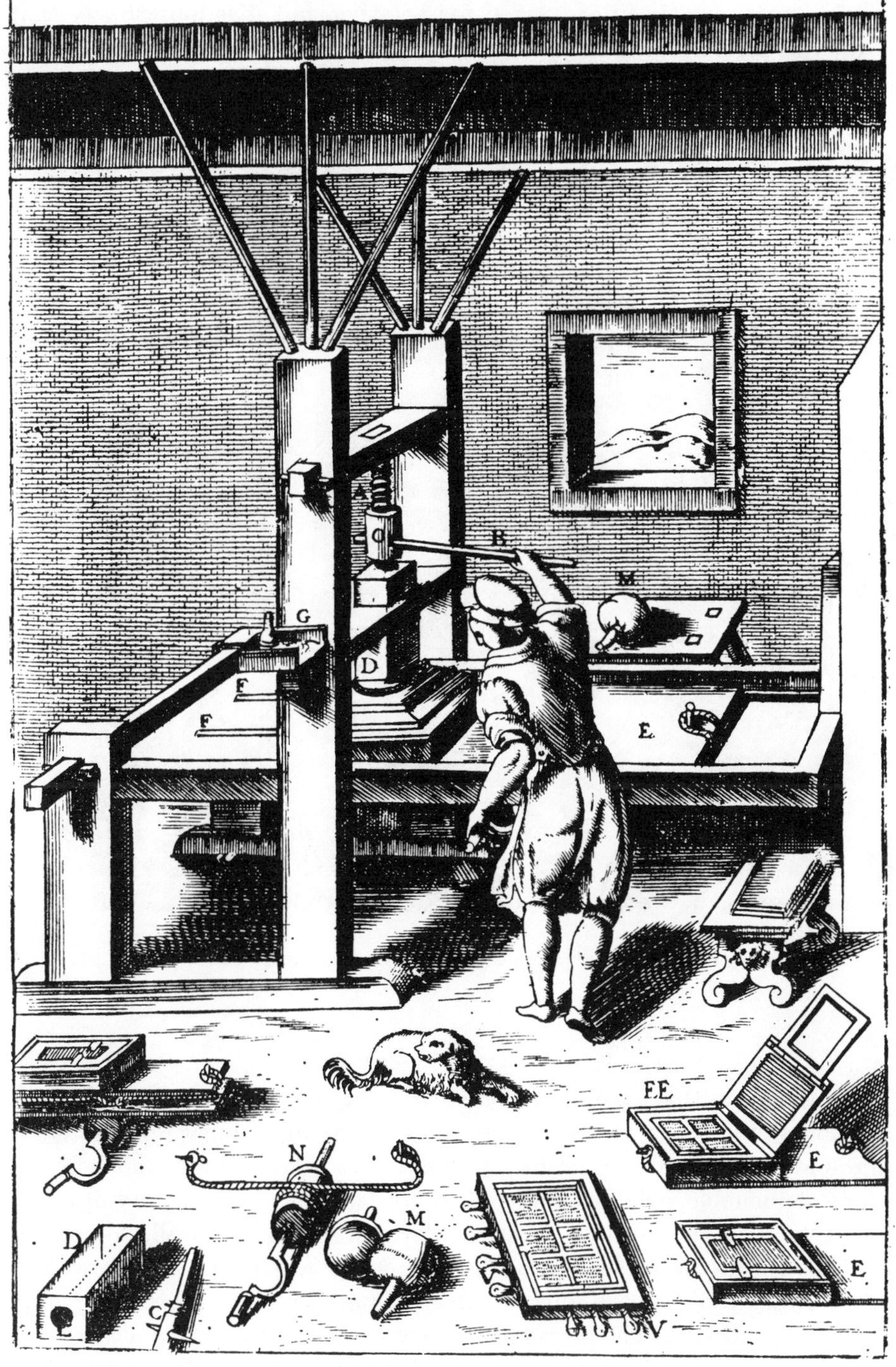

6. Vittorio Zonca's version of the press (1607). (See page 26)

assist in pulling the forme from under the platen. It is just conceivable that an early seventeenth-century printing office was using a press which was out of date a hundred years before, or that, for some economic reason, a screw press from some other trade had been converted for printing, but it is more likely that the artist did not actually draw the press in a printing office, but relied on his memory of other screw presses.

7. Detail of engraving by Johannes Stradanus showing the screw press

To add weight to this possibility there are engravings and paintings of Stradanus which show simple screw presses. There is an engraving of sugar production, in the background of which is a crude heavy press, with a large screw and holed collar. The long detachable pole is being pulled by two men. His engraving of an alchemist's laboratory shows a press, held by stays to the wall, for extracting juices. This engraving served as an inspiration for a painting on the wall of the study of Fransceso I de' Medici, in the Palazzo Vecchio, Florence, completed between 1570 and 1572. The press has a screw working down into a basin which has a channel for the juices to run off. None of these presses would suggest the independent platen of the printing press, and consequently if Stradanus used them as models he would miss this important characteristic.

Stradanus is mentioned particularly because his engraving is one of the most often reproduced to illustrate an early printing office, and seems to have misled generations of artists who have not looked at an actual wooden printing press to observe the working parts, and hence have shown presses which could not have worked.

8. Representation of a press in Haarlem, 1628. Between the tools on the head are the arms of Haarlem

An illustration which is considered to be an accurate representation of a seventeenth-century printing office appears in Peter Scriverius, *Laure-Craus voor Laurens Coster* (Haarlem, 1628). This shows a plain-looking press, but which is technically accurate. From Badius's devices we get the impression that the French printers took a lead in the production of the more elegant-looking presses. In the 1520 device the space between head and cap has been boarded up and terminates in a castellated top. The crude stays to the ceiling have been replaced by adjustable jacks in the shape of pillars. Badius labelled his presses 'Prelum Ascensianum', and in his 1522 device the words are enclosed in a decorative cartouche, while the screw terminates at its top in the head of a cherub. Badius's style was followed by other printers in their devices, some using his wording, while others, such as

Jean le Preux in 1561, substituting 'Typographicum' for 'Ascensianum'. The tradition of using imitations of the arms of Fust and Schoeffer in the colophons of books persisted for many years, and in the *Schweizer Chronik* illustration a shield of Schoeffer's arms hangs on the cheek of the press and his name appears vertically arranged. The splendid 1676 engraving of Abraham von Werdt, of Nuremberg, shows a press, a fine piece of joinery, with a carved griffin on the top of the cap holding ink-balls in its claws. This griffin was the crest on the coat of arms adopted by the German printers' guilds. What existed in reality and what we must attribute to an artist's imagination is difficult to assess, but tastes differed when it came to moulding and carving on presses. There were variations both in structure and decoration. A majority of surviving presses possess head bolts and are undecorated, but as their survival through wars and revolutions (both political and industrial) is fortuitous too much reliance should not be placed on this fact. However, in principle, presses throughout Europe did not differ greatly up to the beginning of the nineteenth century.

We have to wait until 1567 for the first description of the workings of the press, which occurs in *Dialogues françois pour les jeunes Enfans*, published and possibly written by the Antwerp printer Christopher Plantin. Summarized, this explains that the forme, its size depending on that of the page, is handed over to two printers who operate the press. This consists of two cheeks set upright on two feet, which are joined by two cross-pieces, known as 'summers' (head and winter), and is made secure above with stays, pins and keys, which hold it fast and steady. Between the uprights the screw is located, being fitted in the hose, the pivot or head entering a nut, supported by crampons. Its foot rests in a stud bedded into the top of the platen—a large and broad piece of iron—and is attached to it by means of rings. There is a bar which, when pulled, lowers the spindle and, being pushed back, raises it. The forme rests on a stone set in a coffin, at the four corners of which there are corner irons holding the chase. The coffin is on a plank with cramp irons underneath and runs backward and forward along the ribs by means of a spit below, upon which the rounce is fitted. The main structure along which the coffin moves is supported at one end by the press and at the other by a wooden upright (the forestay). On the far end of the coffin is the tympan, made up of two parts, the outer and the inner, between which blankets are placed to avoid the heavy platen battering the type. Paper is laid on top of the tympan, which has two points, fastened by screws and nuts. The paper is pressed on the points. When the paper is turned over for printing on the other side the points go through the holes to provide register. The frisket is then brought down over the paper. The frisket consists of parchment covering all parts of the forme not to be printed. The paper has been dampened the previous day in order that the ink will hold. The type is inked, the tympan lowered, the rounce is operated to bring the coffin half-way under the platen. The bar is pulled once and then the coffin is run in for the other half and the bar pulled a second time. This operation is necessary because the platen cannot cover the whole forme. The coffin is run out, the tympan and frisket lifted and the printed sheet taken off.

This simple description could have been applied to printing at any time from 1500 to 1800 and it also indicates that by Plantin's time the wooden printing press had reached a high point in its development, after which only minor improvements were made. Jost Amman's verses, in his book of trades of 1568, do not add much to our knowledge of the press, but Joseph Moxon's *Mechanick Exercises* (1683–4) not only provides the first detailed description of printing and printing equipment in English, but also indicates what changes had occurred during the seventeenth century.

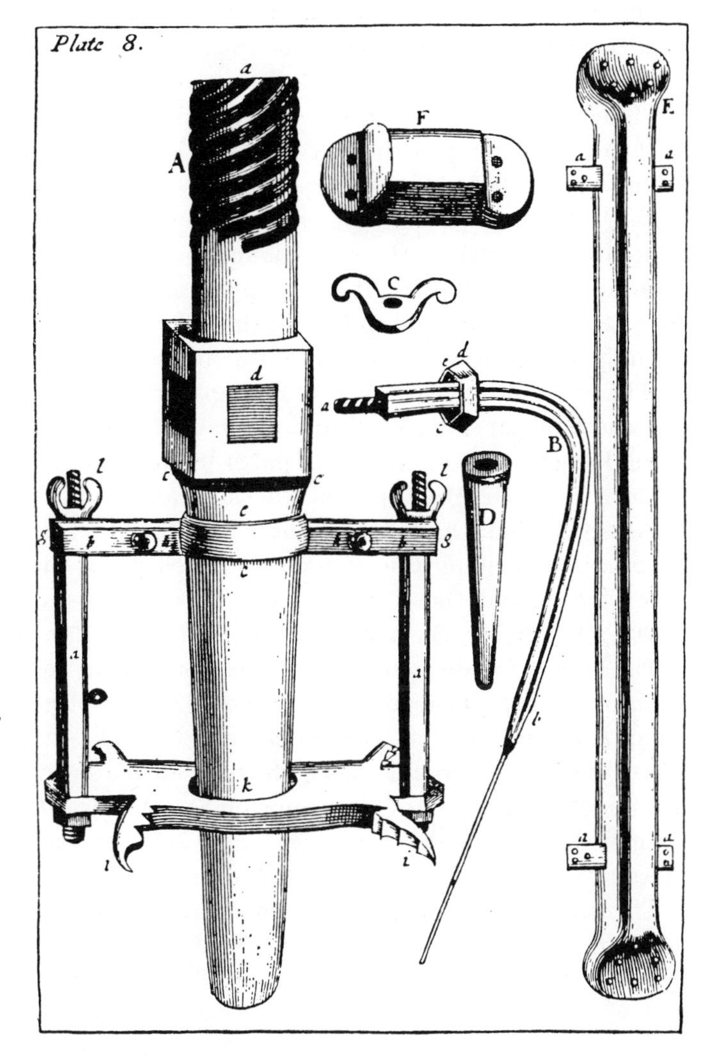

9. A plate from Moxon's *Mechanick Exercises* (1683–4). The spindle (A) works in a 'new-fashioned' or 'Blaeu' hose. The end of the curved bar (B), shown separated from the wooden handle (D) terminates in a male screw. The bar was pushed through the eye of the spindle as far as the shoulder, and then made fast with a wing nut (C) containing a female screw. Two ribs (E) were fixed to the carriage, along which the coffin travelled, assisted by cramp irons (F)

The wooden press became known as the 'common press' if it did not have certain improvements attributed by Moxon, and Moxon only, to William Blaeu (1571–1638). These were a better catch for the bar, improved means of attaching the girths, an adjustable gallows, and gutters to carry off water; but the main feature was the hose which was made of iron instead of wood. The 'English' method was to house the spindle in a long wooden box which passed through a square hole in the till, a cross-piece above the platen. A garter, consisting of two metal half-moons, held the spindle firmly round a groove, just allowing it to rotate but not to twist the platen. The 'Dutch' method was to encircle the groove in the spindle with a metal collar attached to two bars of iron, their own ends being screwed on to a plate, on the four corners of which were hooks for the platen. This stronger hose, ascribed to Blaeu, may have been more efficient in reducing the danger of twisting and slurring the impression, but it was not noticeably adopted in Britain, although, curiously enough, it appears in the illustration of the printing press in the 1787 edition of the *Encyclopaedia Britannica*; and four of the fourteen surviving wooden presses in Britain have modified Blaeu hoses. Using Moxon's description for convenience, hoses can thus be grouped into those of the box type and those of the Blaeu type, although there are variations of both. When presses began to be built in North America the box hose was left open; and with the Blaeu hose four columns instead of two were sometimes used to steady the platen.

The ceiling braces, so important an aspect in early illustrations, seem to have been dispensed with at times. Luckombe in *The History and Art of Printing*, 1771, maintained that the weight of the press alone was sufficient to prevent it being distorted by the constant pulling of the bar.

The power of the wooden press was limited to the amount of force the platen could exert upon an area of type. Perhaps, at the beginning, the platen, type forme and sheet of paper were all approximately of the same size, but when printing more than one page on a sheet of paper became the requirement, the limited power made it necessary to print half a sheet at a time, with two pulls of the bar. Presses were built as large as possible within the limits of the power available. Moxon, in describing the 'improved' Dutch press, mentions a platen as small as 14×9 inches, but those of surviving wooden presses range from $16 \times 9\frac{17}{20}$ inches to $20\frac{3}{4} \times 13\frac{3}{8}$ inches, the average being about 18×12 inches.

The largest wooden press ever built would seem to be that mentioned by John Johnson in *Typographia* (1824)—a double royal size (platen—2 feet 2 inches by 1 foot 8 inches) built for 'the old Duke of Norfolk'. It was worked by a 'fly' at the top of the screw, but this being stolen from the printer who took over the press, Mr. Couchman, senior, of Throgmorton Street, it was henceforward worked 'with a common bar'. The duke also, according to Johnson, had a miniature press. Miniature presses are referred to in Appendix I, devoted to small and amateur presses.

There seems to be agreement among the authorities on the wooden press concerning its speed of operation. The Frankfurt printing ordinances of 1573 laid this down as at about 240 sheets an hour, while Moxon writes of the 'token'—250 sheets an hour, printed on one side by two pressmen. It seems clear, however, that towards the end of a twelve-hour working day the rate would drop, and a more reasonable average figure would be in the region of 200 sheets an hour.

The wooden hand press, as it finally emerged, was the result of small but important improvements on the early versions, all designed to make it more efficient, but none deviating from the basic principle of vertical pressure. The illustration from *Rees's Encyclopaedia* of 1819 may serve as a basis for description of the 'English' press in its final stage.

The two main uprights (the cheeks), about 6 feet high, fitted into the feet at the base and were joined by a beam across the top, known as the cap. The sturdy cross-piece nearer the foot, the winter (*f*), assisted in holding the cheeks together.

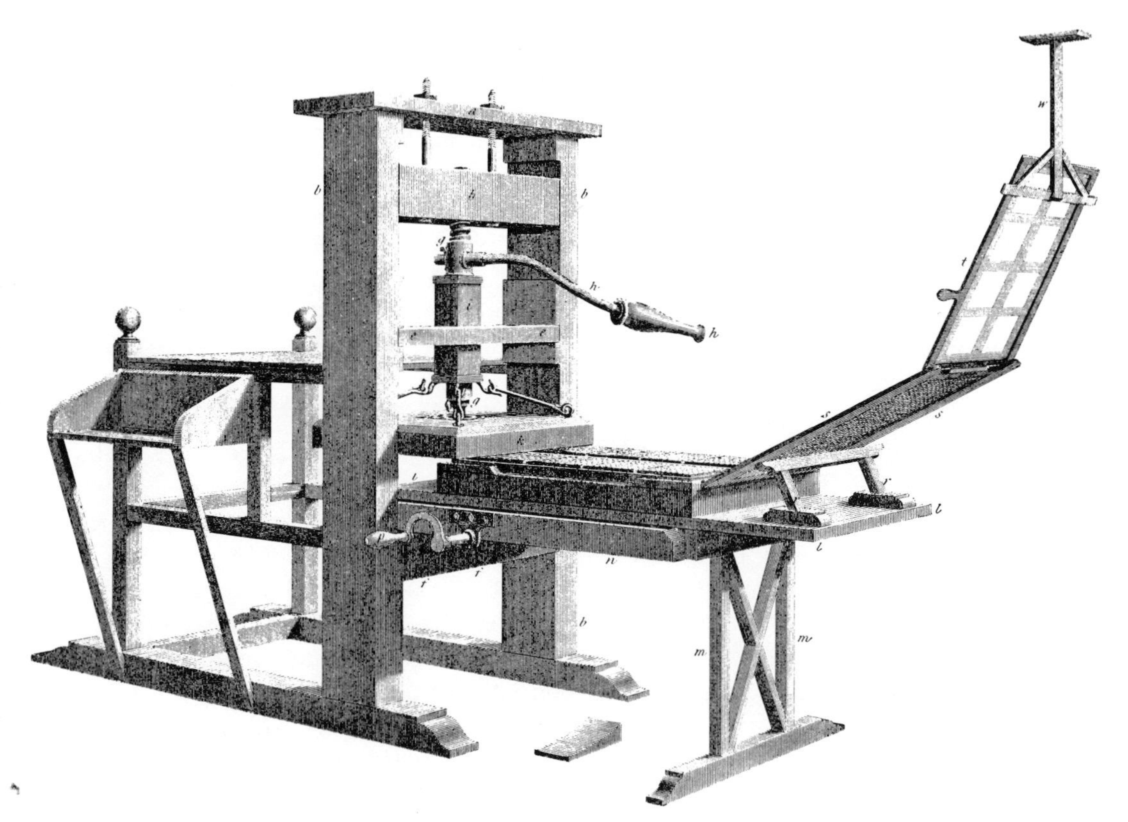

10. Late English press (1819). (See this page)

Not shown is an optional addition, the 'summer', a rail joining the cheeks just above the feet. The head (*b*), the transverse into which a hole was cut to admit a brass nut in which the screw of the spindle worked, was fixed to the cap with two long, adjustable bolts (the head could be raised by screwing up the bolts to determine the strength of pull). The curved bar fitted into the eye of the spindle (*g*), being fastened with an iron key. The spindle then entered the hose, which passed through another cross-piece called the till (*e*). The platen (*k*) was hung from hooks at the base of the hose (usually with strong cords, but here with metal rods). The horizontally moving part of the press, known as the carriage, consisted of a frame (the coffin) fixed to a plank. The type forme was laid on a stone bedded in the coffin. Movement along ribs was by means of a drum and girts, or straps, operated by a device known as the rounce. This horizontal part of the press was held up by a fore-stay.

The tympan (*s*), to which the paper was fixed on points and which was used to control impression, was hinged to the end of the coffin and, when in an upright position, rested against a gallows. The frisket (*t*), a frame covered with a protection sheet with holes to correspond with the type matter, when turned down secured the sheet of paper against the tympan. In the thrown-up position, as shown, it rested against a stay. This was sometimes a rope, or a wooden batten (as here), but Moxon says that a wall could equally perform the office of the stay.

At the other end of the press were the hind posts and, attached to them, the ink block and pegs for the ink-balls. Beneath the press was the foot step for the convenience of the pressman pulling the bar, which when returned was held by a catch fitted to the cheek; but this is often not illustrated.

The procedure followed, the most important part of which became known as 'making ready', constituted much of the 'art and mystery' of printing. The techniques were developed from experience in order to get the best possible result in the time available and with the material supplied. They continued into the era of mechanical printing, especially on cylinder machines, and, to some extent, on rotary presses, although increasingly the improved precision of these later mechanisms exploded much of the 'mystery'.

The paper to be printed had been dampened with water, a few sheets at a time in a trough, and laid on a board with a few dry sheets on top. More wet and dry sheets interlayed had been added until a heap had been prepared to squeeze in a screw press to remove excess water. The degree of damping depended on the nature of the paper and the experience of the printer.

The pressman examined the first impression, which was generally irregular, and the methods he adopted to overcome the defects—the making ready—would indicate his superiority or otherwise as a printer, always bearing in mind the state of his equipment. The process generally consisted of placing pieces of paper below those parts of the type or blocks which were not printing well—called underlaying—to achieve equality of impression. This might be considered sufficient, according to the final effect desired, but since every piece of type might be of a minutely different height to its fellow, not to mention the various parts of a wood block, the further process of overlaying could be carried on to whatever degree was felt necessary. A printed sheet was pulled and the pressman would cut away the parts too heavily impressed, and paste more paper on those which were too light. The finished sheet, known as a skeleton, was placed in the press exactly above the forme. A second sheet could then be pulled to see what progress had been made, and the same procedure adopted. The number of times this was repeated depended on the quality of the work required. The more complicated a woodcut, or later a wood engraving, the more skill was required. The making of skeletons to bring out the full character of an engraving became a highly important task.

The invention of the all-metal, and more precisely built, hand press facilitated the production of fine illustrated works, using the make-ready techniques outlined. At the other end of the scale, a pressman faced with an old, unsteady press and worn types, was tempted to keep make-ready down to a minimum and damp not only the paper but also the tympan, which he would fill with soft blankets, in order that the type might be pressed deep into the paper in an effort to obscure its defects. It is noteworthy, in this connection, that the much-admired printing of John Baskerville of Birmingham was the result not only of his improvement of the press, his specially made paper and ink, but also of his rejection of soft packing in the tympan—the replacement of flannel with layers of superfine cloth, between which was inserted a thin piece of cardboard.

Two men normally worked at the press, taking turns as puller and 'beater' (the one who applied the ink to the forme). The forme having been inked and the paper brought down on to the type, the puller turned the rounce with his left hand, giving about one turn to bring half the carriage under the platen. Placing his right foot on the foot step he grasped the bar with his right hand, sliding it down and giving a straight pull. The spindle turned and forced the platen on to

11. Wood engraving (1676), by Abraham von Werdt, depicting a seventeenth-century South German printing office. The hour glass is used to fix the change-over time of the two pressmen (one beating, the other pulling). The frisket stay in the form of a rope, known in German as the 'Imham', could be ingeniously operated by the puller. At the touch of his foot the rope would stretch and cause the frisket to drop on to the tympan. (See page 30)

the back of the tympan, pressing the paper on to the inked type. The bar was gently returned, the platen rising clear of the tympan, and the rounce was given a further turn bringing the front half of the carriage beneath the platen. A further pull printed the rest of the sheet. The bar was returned, and the carriage run right out so that the paper could be changed and the forme re-inked.

The pressmen might be assisted by a boy known as a 'fly' (not to be confused with the 'fly' of a coining or similar press). He would take the sheets off the tympan as they were printed. Moxon says such an assistant was also known as a 'week-boy'. By his employment production could be speeded up, and the coming of the newspaper led to the more frequent use of these boys. The word 'fly' persists to this day. When Robert Hoe in 1846 marketed an automatic device to convey printed sheets to a delivery table it was known as a 'flyer', and men who take the printed newspapers off a rotary press are known as 'fly-hands'.

12. Badius's 1522 device does not reveal any further technical advances but shows additional decoration (a cherub terminal to the screw) and printers more elaborately dressed than in the 1520 device

The drawbacks of the wooden press, including its lack of stability, and the need for two pulls to print a sheet of paper, had long been apparent. The earliest efforts towards more efficient working included the substitution of metal for wood, experimentation with different metals, and more careful construction of the press by joiner and smith.

While Plantin described the platen as being of iron, he himself had platens made of copper in 1571, and Zonca (1607) recommended a spindle and platen of brass in preference to iron as a metal easier to work. In the Low Countries platens were sometimes fitted with copper or brass plates, and Baskerville ascribed his good presswork to the brass platens made to his own specification. He also, incidentally, used 'Blaeu' hoses on his presses.

Nailing the girts on the plank and coffin damaged the woodwork, and the substitution of clamps and screws not only eliminated the damage, but made the girts easier to adjust. Improved methods of running the carriage in and out were devised. These changes were no doubt helpful, but the improvement so much desired—the printing of a sheet at one pull—was hindered by the lack of power conveyed by the screw and bar, and, ultimately, by the fact that the main construction was of wood.

The need to overcome this restriction became more urgent with the growth of literacy and the demand for more printed matter towards the end of the eighteenth century. A faster-working press was required, but this was not achieved until inventors began working on new principles and were able to harness steam power to printing machines.

13. Sketch by Albrecht Dürer (1511), probably drawn from memory after a visit to the printing office of his godfather, Anton Koberger. The thread of the spindle is incorrectly drawn. If the bar were pulled the platen would move away from the type. The carriage would not come out far enough for the tympan to rise

In the meantime, many efforts were made to produce a 'one-pull' press either of wood, or by making a press completely of metal. Up to 1800 the concentration was on the wooden press with one exception (that of Wilhelm Haas, which will be considered separately); but because of its very nature the wooden press could not provide the best solution. The all-iron press, in its turn, proved to be not much faster in operation than the wooden press, although the printed impression was sharper, and because of its easier working must have made some difference in the number of sheets produced in a day's work.

14. Pierres's 'new press' is illustrated in this portrait engraving of A. F. M. Momoro
(See page 43)

2

IMPROVING THE WOODEN PRESS

THERE WAS HARDLY a time when there was not a feeling that the efficiency of the wooden printing press might not be improved, and almost from its inception, as already indicated, small changes were made. To summarize, the major improvements needed to make the hand press more efficient were greater stability of structure; ability to print a forme at one pull; a reduction in manual effort; and an automatic return of the bar after pulling. Ultimately, all these requirements were achieved at the turn of the eighteenth century in a rigid all-metal press, which had sufficient but easily applied power to print a sheet of paper at one pull.

Leonardo da Vinci (1452–1519) lived during the period when the first printing presses were emerging, and it is said that he contemplated building an improved mechanism. This he is not known to have done but in one of his notebooks (now in the Biblioteca Ambrosiana, Milan) there are two sketches, which show the way his mind was working. One shows simply a screw, bar and platen of a printing press, but the other is of a labour-saving device, by which one pull of the bar would both move the forme under the platen and cause the impression to be made. At the top of the screw is a crown wheel with its teeth working in another at the top of a vertical shaft, parallel to the spindle, and terminating in a grooved wheel around which there are straps attached to the coffin.

No tympan is shown in the plan but sketchily indicated is a sloping bed down which the coffin and forme would slide after printing. The theoretical approach was logical but whether, in practice, the press would have been successful is doubtful. Even if the careful co-ordination and timing could have been achieved so that the forme arrived under the platen just as it was coming down, the effort to achieve this would have added to the pressman's burden, already great, as he would have had additional resistance to overcome. The more effective approach was to increase the power applied to the bar in order that a sheet could be printed at one pull, and to cut down the amount of human effort applied.

However, in 1951, International Business Machines Corporation put on an exhibition of working models from Leonardo's drawings, built by Dr. Roberto A. Guatelli and assistants. These included two models of printing presses but these can now have only a curiosity value.

Before the arrival of the all-metal press many partial efforts were made to improve the printing press, where possible substituting metal for wood. It might be thought that the advances in techniques of iron-working in the second half of the eighteenth century would have stimulated the production of an all-metal press earlier than 1800, but technical development can be restrained by economic and social forces, and this was the case with the printing press. The old wooden press not merely survived well into the era of the metal version, but for reasons of price, prejudice and ease of transport (particularly in the expanding United States of America) wood continued to be a basic element in manufacture. The Belgian Ministry of Industry, in a report as late as 1911, stated that the 'improved' Dutch press (i.e. the so-called *Blaeu* press) was still in use in small rural printshops, notably in Limburg, although only for printing posters—a fate reserved for its descendants, the iron hand presses, which are so used to this day.

In the United States there were a number of efforts to make the wooden press more efficient by, for example, John Goodman, of Philadelphia, in 1786, Belknap and Young, of Boston, in 1796 and Henry Ouram, of Philadelphia, in 1800. These were similar to those which had been tried out in Europe—cutting the screw more evenly, substituting screws for cords to hand the platen and using roller instead of cramp-irons for moving the carriage. Not until 1809 was there an effort in the United States to do away with the conventional screw.

In England, however, in 1771, the Bristol printers and typefounders, Isaac Moore and William Pine, tried a new approach. They substituted for the screw and bar mechanism a platen fixed to a lever of the second order, where the weight is between fulcrum and power. The weight was represented by a connecting rod attached to a platen, which was between the point where pressure was applied to the lever and the fulcrum. They were reverting, in effect, to the beam press which, it is conjectured, Gutenberg had rejected in favour of the screw press. The drawback would have been that Moore and Pine's press could not have produced much power, and was therefore probably restricted to the printing of small cards only. As such, it will be considered in the Appendix devoted to small presses.

A similar press is thought to have been built in 1810 by Benjamin Dearborn of Boston, Massachusetts, but little is known of it, possibly because the simple lever did not provide sufficient power. Compound levers provided the answer, as will be recorded.

Moore and Pine did give some thought to printing a full sheet at one pull, for they patented another press which utilized the 'fly' as used on the hand-operated coin press; in other words, the lever by which the screw was driven was a long metal bar weighted at both ends. The coining press operator, with a circular motion, swings the bar, imparting great momentum to the weights, the effect being to drive down the screw with considerable force.

Whether Moore and Pine ever built this press is not known, but certainly in the next year, 1772, Wilhelm Haas, a typefounder of Basle, was applying the same idea to his newly invented press. He followed the conventional pattern of the existing press, but provided a cast-iron frame, which was mounted on a stone base. The platen and spindle were of iron, and the hose and nut of brass. The platen was depressed by a screw, but its area was greater than that of the wooden press—340 square inches instead of 230. The extra power needed to print this size was provided by the use of the 'fly'—the bar being an adaptation of the lever on the top of

15. Seventeenth-century screwmakers at work. In the foreground can be seen the frame of a coining press, adapted by Wilhelm Haas to the printing press. (See this page, page 32 (Duke of Norfolk) and page 82 (Russian press))

a coining-press screw, extended both sides of the spindle and loaded at each end with a weight. When the bar was pulled by the pressman the same effect was obtained as when an operator swung the loaded lever of a coin press. Energy was stored in the moving weights, which, together with the pull, produced sufficient power to print from the large forme.

Haas was opposed by the Basle printers' guild which would not allow him to work as a printer as well as a typefounder and so, for the time being, development languished. But later, his son, another Wilhelm, set up a private printing office, and an improved version of the press was made in at least sufficient quantity to supply a German printer who was impressed by its capabilities. Despite the fact that the staple was made of iron, it tended to break, and a model with a strengthened frame was constructed in 1784. This fracturing of the frame, which later was to concern the inventor of the first all-metal press, Lord Stanhope, indicated that the simple replacement of wood by iron in itself was not sufficient to counterbalance the stress provided by more powerful printing mechanisms. Attention had to be given to the precise design of the iron frame as well.

The *Haas* press represented a great step forward, but it was not immediately followed up, innovators continuing to use the wooden press as a basis for improvement. Among them were members of the Didot family, of Paris, who experimented with a number of single-pull presses during the period 1777–84. Those concerned were the brothers François-Ambroise (1730–1804) and Pierre-François (1732–93) and François-Ambroise's son, Pierre (1761–1853). The elder Didot is reported to have designed a single-pull press which was used to print an edition of *Daphnide et Chloë* in 1778. In 1784 Pierre published a poem which stated that his father's

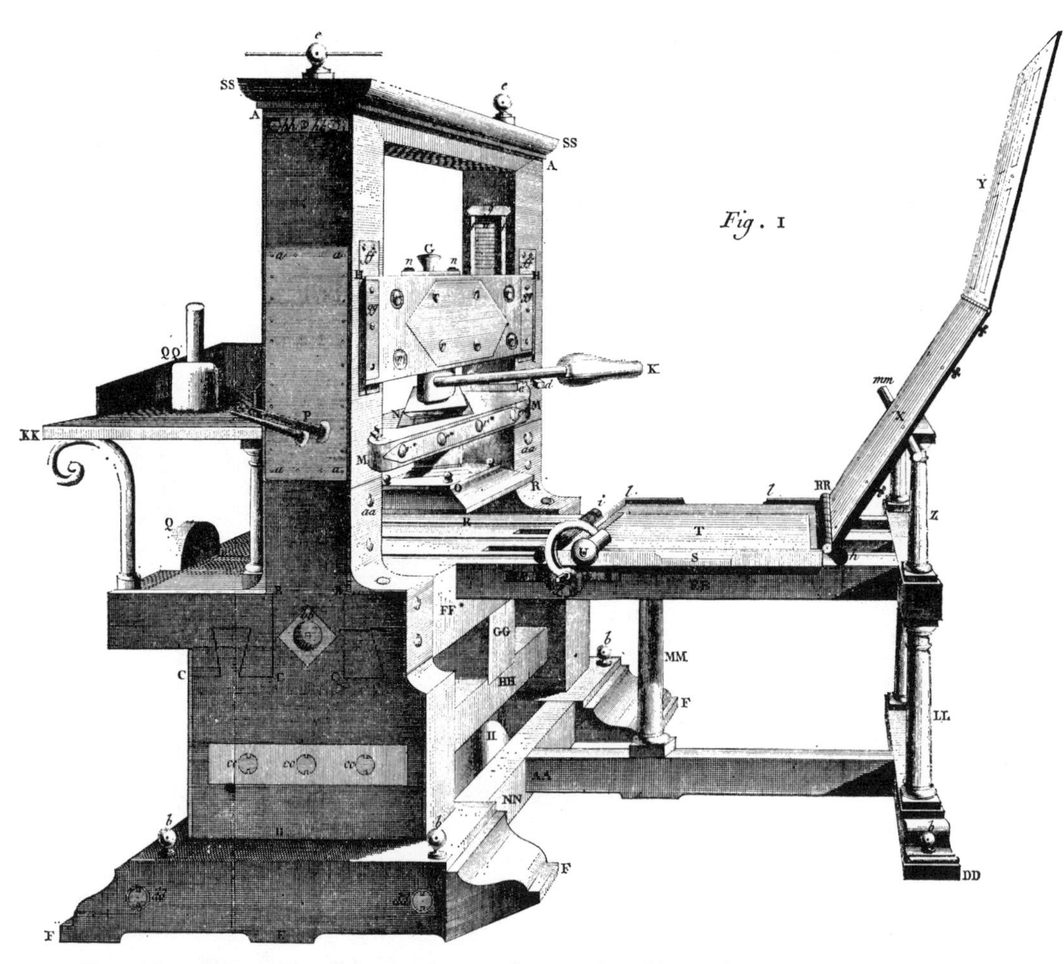

16. Engraving of the Anisson-Duperron improved press. (See this page)

idea had been stolen by another printer; in an edition of 1786 the printer is named as E. A. J. Anisson-Duperron, Director of the Imprimerie Royale. The reference is to two presses, and it is inferred that Pierre assisted his father with both. It was the second, apparently, on which Anisson fixed his attention.

Anisson's mechanism, whether original or stolen from Didot, was basically a common wooden press, although of a much more solid structure. His important contribution related to the screw, which had two threads. The first, or upper, worked in the head in the usual way, lowering the spindle a short distance; the second, or lower thread, ran in the same direction as the upper but its pitch was not so steep. Operating the bar resulted in a very small descent in relation to the press as a whole, but developing more than double the power of the common screw press. This development became known as the Hunterian screw, the mechanical power having the same as that of a single screw with a thread whose magnitude is equal to the difference of magnitude of the two threads.

An *Anisson* press was officially installed in the Imprimerie Royale in 1783, and remains today in the now Imprimerie Nationale, but whether more than one was built is not known (Plate III). Anisson was guillotined in 1794.

Philippe-Denis Pierres (1741–1808), printer in ordinary to the French king, invented a number of printing presses, the best known about 1784. This had a platen twice the size of that of the common press, being $25\frac{1}{2}\times19\frac{1}{4}$ inches, and utilized a cam movement instead of a screw. It also had a bar which pushed down vertically instead of being pulled horizontally. Pierres's press was therefore of a distinctly unusual nature, being the first effective large press to dispense with the screw. The cam operated between bearings set under the head and over the top of a piston which penetrated the till, terminating in a platen held in an adjustable box. The cam was fitted to a spindle fixed to a wooden quadrant, round the circumference of which ran a chain, fixed at one end to the top of the quadrant and at the other to a lever pivoted on the back part of the press. When the lever was pressed it pulled the chain down, turned the quadrant, and thus the cam, which applied vertical pressure to the piston. After impression the platen was raised by means of a spring above the till, while the quadrant was returned to its former position by a weight and levers. Pierres published his *Description d'une nouvelle Presse d'Imprimerie* in Paris in 1786, and since the imprint states that the book was printed on the new press, one model, at least, must have been built. An engraving of A. F. M. Momoro, 'Imprimeur de la Liberté Nationale', dated 1789, has a small illustration of Pierres's press in the corner.

An employee of Pierres, Genard, attempted to improve the press, and a model survives in Paris in which the only significant change is that a screw is substituted for Pierres's cam. Genard actually put the improved press on the market, and it may be that the *Apollo* press exported to England before the end of the eighteenth century, and mentioned by both Johnson and Hansard in each version of *Typographia*, was a development of the Pierres press. Johnson wrote: '. . . about half a dozen beautiful presses upon the French construction, with long levers, which were pressed downwards, were in the possession of three or four Master Printers of the Metropolis.' Hansard added that as the long lever worked up and down like the lever of a pump 'requiring a motion of the whole body was found hurtful to the men, and in consequence they were soon disused'.

From Isaiah Thomas's *The History of Printing in America* (1810) we learn that in about 1785 Benjamin Dearborn, a printer of Portsmouth, New Hampshire, and subsequently of Boston, constructed a 'wheel press', which printed the whole side of a sheet at one pull of the lever. The platen turned with the tympan, having a counterpoise to balance it, and the power of the lever had the additional force of a wheel and axle. The press was used by John Mycall, a printer of Newburyport, Mass., but after one or two efforts more to improve printing presses, Dearborn went on to other, and presumably more profitable, fields of endeavour.

In London, Thomas Prosser patented on 9 December 1794 a converted wooden press which had two springs—one above the head and the other below the winter or base. Each spring had a regulator and by moving these nearer to or further away from the centre of the head of the press the pull was made harder or softer. Prosser claimed that the impression was materially increased and that the labour applied was considerably reduced, but there is no record of the workings of this press, except that Hansard denounced it as 'making bad worse'.

During the next year the Society for the Encouragement of Arts, Manufactures and Commerce awarded Joseph Ridley forty guineas for his suggested improvement of the printing press. Ridley abandoned the screw and replaced it with a vertical steel bar with a conical end resting in the cup of the platen. Between head and till there ran a horizontal thick iron rod penetrating each cheek, with an upright lever on one end to turn the rod. Three chains were wrapped round the centre of the rod, the two outer ones serving to pull down the platen, and the centre one to raise it. At the other end of the rod protruding from the cheek was another lever with a weight on the end, which, acting as a fly, could be used to give additional power to the impression. Ridley's press was meant to be a single-pull machine and a model was made, the present whereabouts of which are not known. But it is doubtful whether a full-scale press was built as a general impression is gained from such writers as Hansard and Bennet Woodcroft, compiler of the *Abridgment of Specifications Relating to Printing* (Patent Office, 1859), that the *Apollo*, imported from France, was the first press in Britain to print a whole forme at one pull; and that this was followed by Roworth's invention and then by Stanhope's.

As the Stanhope appeared in about 1800, the Roworth must have been earlier. Roworth, a printer of Bellyard, Fleet Street, London, produced a common press but with a plain spindle, working at the upper end in a socket at the head of the press, and at its lower end in the usual cup on the upper side of the platen. Under the socket was a circular collar above which were two circular bands or sectors of steel, creating short inclined planes on opposite sides of the spindle. Above the collar were two studs which, when the bar was pulled, slid along the angles of inclination, causing the spindle to descend, rapidly at first, but with decreasing velocity as pressure was increased. The power increased as resistance increased. The platen was returned by the use of a weighted lever.

Hansard says that Taylor and Martineau, of City Road, London, had a *Russel* press which incorporated a combination of inclined planes or wedges acted on by bars with hinged joints, and a coupling bar working in projections on the staple.

This idea of a wedge and inclined planes as a basis for a pressing device was adopted in 1809 by Samuel Fairlamb, of Marietta, Ohio. A notice in the Albany *Gazette* stated: 'Instead of a screw he substitutes two inclined planes, moving on rollers placed in the summer, in a circle of six inches diameter. The advantage of this invention consists in avoiding the friction common to a screw, and making the pressing power much easier and more effectual. The platen is supported by two spiral springs, which supply the place of a box encircling the spindle. Some minor improvements are added to the above. A press of this construction can be made for about half the price of the common press.'

The principle of inclined planes was also used by George Clymer, of Philadelphia, in 1811, and by Elihu Hotchkiss, of Brattleboro, Vermont, in 1817. Little is known of any of these presses which, considering the period, must have been basically wooden constructions, but there are interesting coincidences which may suggest that one of them may have been transformed into a metal construction in England in 1821 by David Barclay—a point which will be pursued in Chapter 5.

In Britain Roworth was followed by Augustus Frederick de Heine who, in his specification of 1 February 1810, stated: 'Instead of applying a screw for my power I apply two sectors, or a sector and cylinder, or a sector and roller, to move one against the other by a single or compound lever.' De Heine's ideas were merged into the press made by T. Cogger which, as it comes into the period of iron hand presses, as does the *Columbian* which Clymer developed after his wooden presses, will be covered in appropriate chapters.

Another inventor, John Brown, a London stationer, took out two patents, which reveal the kind of activity which was going on at this time to make the printing press a more efficient and speedier instrument. In Brown's first patent of June 1807 the platen was drawn down 'by a screw turned by a bent lever acted on by a cord passing round a rigger on an upright shaft, to which motion is given by bevel gearing from a winch'. After the impression, the rigger was thrown out of gear and the platen recovered itself by means of a weight or spring. But perhaps the most significant aspect was the proposal to partially mechanize the inking of the forme, which seems to establish Brown as a pioneer—he claims that his 'inking apparatus' is his principal improvement. But he was not the first to attempt the substitution of the ink-ball with a roller. This had been attempted by Lord Stanhope by at least 1803. Brown's patent refers to a cast-iron bed carrying the forme sliding out below an inking roller, which 'is covered with flannel, or any other proper elastic substance, and then is covered with parchment or vellum, or other proper materials, to prevent the ink from soaking too far in, and likewise to give it spring, and afterwards is covered with superfine woollen cloth for the purpose of receiving ink to supply the types'.

A large barrel or cylinder having received the ink from the trough underneath it, had it distributed by a small roller in contact with it, supplying the main ink roller, and the latter 'revolves round and feeds the types by the motion or movement of the spindle which moves the bed'. Brown also envisaged driving the inking apparatus by means of a fly-wheel and 'traddle'. Here he was approaching the idea which was to be used on the jobbing platen presses of three decades later.

Brown's patent of November 1809 was more ambitious still. The forme is fixed in the centre of the press. An upper bed is rigidly connected with a carriage which runs on a tramway above the press, and is drawn over the forme by a cord passing over a roller turned by a hand lever. The impression is given in three possible ways. The first by the rising of a vertical rack below the bed by motion imparted through a series of levers from a rounce barrel; the second by a screw acting upwards; and the third by a roller fixed on the bottom part of the carriage 'where the upper bed is, and by the same motion rolls over the tympan and types and gives the impression'.

To increase production Brown proposed the use of two tympans, one on each side, so that two workmen might use the same forme. His inking apparatus consisted of three horizontal endless webs placed one above the other, the lower one being above a trough roller made of metal and hollow inside so as to contain a quantity of hot water or steam to keep the ink of a proper thickness. The action of his apparatus was complicated, but it seems that the lower web was furnished by the trough roller by being raised by treadle action, and would convey the ink to the upper rollers. Whether all this would have worked or not it is difficult to estimate, but his rollers sewed round with leather or cloth, would have had the same drawbacks as those of Lord Stanhope, and later of Friederich Koenig, inventor of the first cylinder press, in that the seam caused the inking to miss. The solution to the problem, discovered in time for Koenig to take advantage of it, was to cast a smooth roller from some composition material. Nothing is known of the fate of Brown's experiments, but he was very much ahead of his time, and his idea of double tympans so that two men could work the same forme was later taken up by others.

Another modified wooden press, using a different principle, which carried over into the iron-press era, was that of George Medhurst (1759–1827), clockmaker turned ironfounder and scale-maker, of Soho, London. Instead of a screw, a plain spindle was employed, on the lower part of which, just above the lever, a circular plate was fixed. From the plate to the head there extended two iron rods, which, when the platen was up, were in an inclined position. When the bar was pulled, the circular plate turned, the lower parts of the rods turning with it. As the upper ends of the rods were fixed, the effect was to bring them into a vertical position, forcing the platen to descend. According to Hansard, the power increased as the resistance increased, and when the rods came nearly parallel to the spindle, or into a vertical position, the power was 'immensely great'. This is an early use in printing presses of a so-called 'toggle', which is frequently met with after 1816. It refers to a mechanism in which pressure is derived from the movement of rods and levers. Medhurst's version is known as 'torsion toggle' from its twisting action.

In their attempts to improve the hand printing press, inventors had to temper technical innovations to the economic situation of the printing trade. It is not surprising therefore that they continued to use the wooden frame, and concentrate on improving the printing mechanism, long after it had been demonstrated, partly by Haas and completely by Lord Stanhope, that a metal structure was desirable. The expense of the *Stanhope* press was a limiting factor in its adoption, and C. Stower in his *Printers' Grammar* (1808) accordingly reported that the application of Stanhope's lever to the common press had been attempted 'by nearly all the press-makers of the metropolis', but that experience had shown 'how impossible it is to construct a machine of this sort composed principally of wood, capable of resisting the great increased power intended to be produced by the new arrangement of the bar and spindle'.

The wooden frame, in other words, broke under the stress of the powerful new lever system, although, as it happens, an undamaged wooden press, equipped with 'Stanhopean' levers, survives in the St. Bride Institute collection.

The drawback could not have applied either to mechanisms such as that envisaged by Medhurst, as his idea of inclined rods was the basis of a combination wood and iron press invented by Abraham Stansbury, of New York, in 1821. That the power could not have been all that great is evident, as the first Stansbury with wooden cheeks was a two-pull press, with a bed twice the size of the platen. An all-iron *Stansbury* was later produced, being manufactured into the last quarter of the nineteenth century. Another press which utilized the inclined rods was the German-built Hagar.

Rarely were specific improved wooden presses manufactured in any quantity, but one group did achieve the distinction of a trade name—those manufactured to the number of 1,250 by Adam Ramage, a Scot who emigrated to the United States in 1795. As with other later presses, the name 'Ramage' in fact became a generic name, and has been applied to surviving early presses not of Ramage's manufacture.

In 1799 Ramage was named in the Philadelphia City Directory as a printer's joiner, and other records indicate that he repaired common presses. From about 1800 he went on to manufacture presses and from 1807 to make his own modifications. At first he enlarged the screw and reduced the pitch, thus doubling the power of impression. The platen was faced with brass, and it was fastened with four bolts instead of cord lashings. His second press, part iron and part wood, dispensed with the cap and head, combining them into one cross-piece, and the hose by using springs on two sides to lift the platen. Instead of the old stone bed he used one of iron and by 1820 was making his platens completely of iron (Plate V).

Ramage's proof press was a smaller, table-mounted version of his screw press, again part wood and part iron, and with a platen size of $12\frac{1}{4} \times 6\frac{1}{2}$ inches.

Ramage was a general press maker, manufacturing not only bookbinding and paper-cutting presses, but also an all-iron frame table press complete with screw, iron platen and bed. He called it the printing, copying and seal press. One survives in the Museum of the Oregon Historical Society, Portland, Oregon. It is a foot high, with a platen of 18×13 inches, and was designed to be a small hand card press, and consequently a tympan and frisket were added. Nevertheless, it was used for larger productions than cards, first by American missionaries in Honolulu, and then, from 1839, in the Oregon country by the Revd. H. H. Spalding at the Lapwai Mission (now Idaho) for educational work with the Indians. A primer and translation of the Gospel of Matthew, both in Nez Pearce, were among items published from the press. In 1846 the press was removed to the Dalles Mission where it was used by Marcus Whitman until the 1847 massacre by Indians at the Mission, when it was taken by the Revd. J. S. Griffin to Hillsboro for the publication of a small bi-monthly paper, the *Oregon American and Evangelical Unionist.*

About eighteen other *Ramage* presses survive in various parts of the United States, with platen sizes of $20\frac{3}{4} \times 13\frac{1}{8}$ and $19 \times 12\frac{1}{4}$ inches. Testimony to the service of Ramage wooden presses lies in the fact that they were favoured by those setting up as printers in the expanding parts of the nation, as they were lighter and easier to transport than iron presses, which at first were not only heavy but bulky. This fault was rectified later when lighter iron presses with detachable parts were produced. The first printing press in New Mexico was a *Ramage*, royal size, brought over the Santa Fé Trail by wagon from the United States in 1834, and that of California followed soon after by ship. Like those of New Mexico and California the first printing presses in Utah (1849), the State of Washington (1852) and possibly Colorado (1859) were *Ramages.*

The first half of the nineteenth century saw a proliferation of iron hand presses, all continuing to use the flat platen, but with a variety of devices to drive it downwards. However, the first of their number retained the use of the screw. It was the *Stanhope* press, the subject of the next chapter.

17. A Stanhope press of the first construction at work. This, the first all-iron printing press, was invented in about 1800 by Earl Stanhope. From about 1806 onwards the press was improved with strengthened cheeks. Those printers who had purchased the first model did not dispense with its services even if the sides tended to crack. They had the sides repaired. The printer of the book, in which this engraving appeared in 1831, was C. Whiting, of the Strand, London, who may have had one of the early models

3

THE STANHOPE PRESS

BY THE END of the eighteenth century the time was ripe for a major step forward in printing-press construction, and the reason why this was possible in England was because of the advances which had been made in the techniques of casting metal, and in the rise of a class of mechanics, the forerunners of the engineers, who were to transform the nineteenth-century industrial scene. That the man who grasped these facts and used them to produce the first all-metal press was not a tradesman but a peer of the realm was not surprising. The third Earl Stanhope (1753–1816), who was devoted to scientific inquiry, was free from the conservatism of the average printer and had greater resources at his disposal.

Wilhelm Haas, of Basle, had been the first to use a cast-iron staple for a hand press, but increased the power by use of a 'fly' or weighted lever, one criticism of which was that it increased the danger of slurring the impression. When Earl Stanhope came to invent the press which bears his name in about 1800, he adopted a different approach. Like Haas he retained the conventional screw but separated it from spindle and bar, inserting a system of compound levers between them. The effect of several levers acting upon another is to multiply considerably the power applied. As a result, in the *Stanhope* press the descent and ascent of the platen were accelerated, and the power was gradually increased until it was greatest when forme and platen came into contact.

The platen was of double the ordinary size and this, with the increased power, allowed the forme to be printed at one pull. Stanhope was aware that the power was so much greater than that previously supplied and considered it proper to regulate it to prevent it harming the press. This was affected by a regulator, which was placed against the stop of the upper lever bar. Detailed instructions were issued in July 1805 on the use of the regulator, which, from being a small piece of iron became an adjustable screw at one end of the link between the two levers.

Almost inevitably, Stanhope has been accused of copying Haas, particularly as he had been in Switzerland, although in Geneva, returning to England in 1774; and because of the resemblance of the iron staples, not to mention a counterweight on both presses. It is not necessary to pursue the details of the charge since it is easily answered by reference to the outstanding feature of the *Stanhope* press. Both Haas and Stanhope followed the basic design of the wooden press, attempting no radical departure, but Stanhope made an original contribution with his system of levers, which was certainly not copied from Haas.

While the various mechanisms used previously to replace the screw may have been superior in some ways, they were costly and difficult to repair. Stanhope preferred to concentrate on the use of compound levers, on 'Stanhopean principles', as they were called, and which served, being carried over into the *Columbian* press, as a link between the screw and the knuckle, chill or toggle joint, the important characteristic of the *Albion, Washington* and similar hand presses of the nineteenth century.

On a pamphlet dated 1807, entitled *Specifications respecting Ships and Vessels,* by Charles, Earl Stanhope, the imprint is 'London: Stereotyped by A. Wilson, Duke Street, Lincoln's Inn Fields and printed by him in Wild Court, at the Iron Press of the Second Construction invented by Lord Stanhope', from which we learn that there were presses of the 'first construction' and these were those which Hansard complained, in 1825, were liable to 'serious injury'. These presses had a straight-sided frame and, like some wooden presses, were sometimes too weak to withstand the pressure they were called on to bear. Hansard also complained that the screw and box, which contained the nut, being of cast iron, were liable to break, and adds that he had had them replaced by a wrought-iron screw and bell-metal box. He did not like the gallows either, finding it inconvenient and dangerous. Similar criticisms were made by Johnson, but, in fact, each fault was eventually remedied and it is odd that Hansard ignored the improvements and reproduced in his *Typographia* a *Stanhope* press of the first construction which had become out of date some eighteen years earlier. Johnson, at least, in 1824, illustrated a strengthened press with rounded cheeks.

18. Stanhope press of the second construction (1824)

Stanhope would not patent the press himself, and engaged with Robert Walker, an ironsmith, of Vine Street, Piccadilly, London, and then of Dean Street, Soho, to manufacture it. It is known what the earliest *Stanhope* press was like from a photograph reproduced in the *British and Colonial Printer and Stationer*, of 13 December 1906. This showed the first model, which was used by the printer William Bulmer, who was succeeded in 1854 by Nichols & Sons. An effort has been made to trace this press, but without success. Nichols's successors were absorbed by His Majesty's Stationery Office in 1939, and although older members of the staff claim to remember the press there is no trace of it now. In his monograph, *Charles, Earl Stanhope and the Oxford University Press* (1896), Horace Hart, then University Printer, stated that a *Stanhope* press of the first construction (No. 13) was still in existence. There is no trace of it today. Hart did, fortunately, reproduce a photograph.

There are at least three surviving early *Stanhope* presses. One (No. 5, of 1804), originally in use at the Chiswick Press, Tooks Court, Chancery Lane, is now in the Gunnersbury Museum. The other two are owned by newspapers (Plates IX and X) —the first (No. 67), about the origin of which little is known, by the *North Eastern Evening Gazette*, Middlesbrough; although significantly it does have reinforced sides. The other (No. 9) is owned by the *Morpeth Herald*, which was founded in 1854, and first printed on this press, which was probably purchased from some other source in Northumberland.

A *Stanhope* press consists of a massive cast-iron frame formed in one piece, in the upper part of which a nut is fixed for the reception of the screw, the point of which operates on the upper end of a slider, which is fitted into a dovetail groove formed between two vertical bars of the frame. The slider has a heavy platen attached to its lower end, and the weight of the platen and slider are counterbalanced by a heavy weight behind the press, suspended on a lever. The iron carriage is moved in the same way as on a wooden press. The screw is motivated by a bar attached to multiple levers, the important Stanhope innovation, and the whole mechanism is mounted on a characteristic heavy T-shaped wooden base. The original Walker presses are generally inscribed 'Stanhope Invenit / Walker Fecit'; others bear the maker's name and address, John Brook (or Brooks) following the tradition by inscribing 'I Brook fecit'. This manufacturer, of 14 Wild Court, Lincoln's Inn Fields, was originally one of Walker's employees. At least one surviving press bears no inscription at all.

From the photographs, surviving presses and published material, it is possible to see that the design of the press proceeded in stages. The Bulmer press was not made completely of metal, the iron staple being mounted on a wooden frame. The full-size platen, the levers and counterweight are identical with those of the earliest surviving press of 1804.

Various improvements were made, the most important of which was the strengthening of the staple from about 1806 onwards, which gave rise to the characteristic rounded cheeks of the typical *Stanhope* press. According to Johnson the rounce handle was 'attached to a rod which crossed the platten'—a rather cumbersome device which can be seen in Hart's reproduction. Also visible in the picture is the carriage, which ran on wheels and which, Johnson complained, made a very disagreeable noise. The older, well-tried cramp-irons running, however, in troughs, were substituted for the wheels, and the traditional type of rounce restored. Hansard's complaint about the gallows was answered, for it was abandoned when it was found that the arm of the counterweight of the tympan frame did its work. A comparison between the illustrations in Stower's *Printers' Grammar* (1808) and Johnson's *Typographia* (1824) illustrates the point clearly.

Hansard's idea of using wrought iron for the screw was eventually taken up, but not by Walker, who is thought to have died before 1813. He was succeeded by S. Walker & Co., which was run by his widow, Sarah. She, in turn, was succeeded by her son-in-law, S. J. Spiers, in about 1833, so that a printed circular issued by S. Walker & Co. must be before that date. The significance of the circular is in that wording which refers to improvements—'strengthening the staple so as to preclude the possibility of its breaking; cutting the screw in the eye of the staple instead of having a cast-iron box; using a wrought-iron screw instead of a cast-iron one, with several other minor alterations, by means of which the STANHOPE PRESS is not only made much stronger, but likewise lighter, thereby enabling us to sell them considerably cheaper than formerly'.

As to the price of a *Stanhope* press, this was originally ninety guineas—another source of complaint by Hansard—and which was considered to be a check on sales. Stanhope answered this by pointing out that since £60 or £70 was being paid in London 'for presses on the French construction' (that is, wooden presses), the price of the *Stanhope* could not be considered extravagant. In any case, when competitors to Walker entered the market, the price began to drop. Stower (1808) lists prices as:

		£	*s.*	*d.*
Common-size Printing Press		31	10	0
Foolscap	ditto	21	0	0
Stanhope ditto by Walker		73	10	0
Ditto by Shield & Co.		63	0	0
Brooke's	ditto	42	0	0

The 'Brooke's' refers to a wooden press with 'Stanhopean' levers, made by Brooke, of Holborn, London. On the circular put out by S. Walker & Co., mentioned earlier, prices have been added in manuscript. They range from £63 for the Royal size to £70 for Double Crown. No price is given for the smallest (foolscap: platen 1 foot by $1\frac{1}{2}$ inch by 1 foot 5 inches) or the largest (Newspaper: platen 3 feet by 1 foot $11\frac{1}{2}$ inches).

Hansard states that Fowler, Neal, Jones & Co., of the King's Arms Iron Works, Lambeth, made *Stanhope* presses for sixty guineas. Other British manufacturers as well as Brook, or Brooks, were Peter Keir, of Camden Town, who made a number of improvements; D. Saunders, of Cambridge; and Shield & Co., of High Holborn.

Whether this last-named firm had any connection with Francis Shield, of London, who established a printing-press manufactory in New York in 1811, is not clear. The *Long Island Star*, printed at Brooklyn by Alden Spooner, which reported on Shield's activities, is also ambiguous about the precise nature of the presses made by him. Shield had made two presses since his arrival in America, one of which was in the office of the *Long Island Star*, it is stated. Then follows a description of Earl Stanhope's invention and 'we have never heard of more than two of these presses being brought into the United States, one of which is now owned by Messrs. Bruce of New York'. The implication of the report is that Shield manufactured *Stanhope* presses in America.

The situation in Germany and France is clearer. Koenig sent a *Stanhope* to Germany in 1815 at the request of Georg Decker, of Berlin, and a press purchased by Brönner, of Frankfurt, in 1819, was used as a pattern for a press made by the firm of Bayerhofer, and, in 1825, the firm of Scheggenburger made the first *Stanhope* in Berlin. Christian Dingler, of Zweibrücken, made a version, introducing a forestay in the shape of a lyre. Dingler, a prolific adapter, also used the *Stanhope* wooden T-base for his *Columbian.*

As soon as the Napoleonic wars were over Ambroise Firmin Didot came to London and, seeing the *Stanhope* press at work, eventually bought one for his own office in Paris, later importing a *Columbian* as well. Inspired by the *Stanhope,* Parisian press-makers copied it. A. Frey, in his *Manuel français de typographie* (Paris, 1835), lists Bresson, Misselbach, Thonnelier, Giroudot, Frapié, Gaveaux, Durand and Colliot. Others, not listed by Frey, were Coisne, Rousselet and Tissier. Of the French *Stanhopes* Bresson's were said to be the most esteemed. Gaveaux indulged in slight decoration, adding a vase to the top of the screw, and repeating the motif for the forestay. Frapié made a number of changes, replacing the screw with an inclined plane, and providing pressure from below which pushed up the carriage on to a fixed platen. The result was a press which bore the least resemblance among the French productions to the *Stanhope.* A table model *Stanhope* with a name-plate 'La Typote' is described in Appendix I.

A French-built *Stanhope* is still in regular use for the over-printing of bullfight posters at Imprenta Castellanos, in the town of Alcázar de San Juan (Ciudad Real), Spain (Plate XI). With a platen size of 80 × 60 cm., the press carries a brass plate with the inscription Tissier & Cie / Mechaniciens, Rue du Chaulie 16, A Paris / 1847. Imprenta Castellanos was founded by the present proprietor's great-grandfather, José Castellanos, in 1847 in the same building as it is now. The *Tissier* press was bought new and was, in fact, the firm's first press. This is an unusual example of a 125-year-old press occupying only one location in its lifetime.

In Italy the *Stanhope* press was manufactured by G. B. Paravia, of Turin, and Dell'Orio of Monza. A *Stanhope* was still being illustrated in a Russian printing manual of 1874.

The first *Stanhope* press in Sweden was imported from France by P. A. Norstedt in 1828; and a Swedish-built version was first manufactured by T. Munktell, a surviving example being at Skansen, the Swedish craft museum (Plate XII).

While the rate of printing with the *Stanhope* press remained at only about 250 impressions an hour, the advantages over the wooden press were considerable, particularly its ability to print a sheet at one pull.

Hansard gives an interesting description of the difference between working a wooden and an iron press: 'The advantages of the iron presses in working are very considerable, both in saving labour and time. The first arises from the beautiful contrivance of the levers, the power of the press being almost incalculable at the moment of producing the impression and this is not attended with a correspondent loss of time, as is the case in all other mechanical powers, because the power is only exerted at the moment of pressure, being before that adapted to bring down the platten as quickly as possible. In the Stanhope press, the whole surface is printed at once, with far less power upon the lever than the old press, when printing but half the surface. This arises not only from the levers but from the iron framing of the press, which will not admit any yielding, as the wood always does, and indeed is intended to, the head being often packed up with elastic substances, such as paste-board, or even cork. In this case much power is lost, for in an elastic press the pressure is gained by screwing or straining the parts up to a certain degree of tension, and the effort to return reduces the pressure: now in this case the handle will make a considerable effort to return, which, though it is in reality giving back to the workman a portion of the power he exerted on the press, is only an additional labour, as it obliges him to bear the strain a longer time than he otherwise would. The iron has very little elasticity, and those who use such presses find it advantageous to diminish the thickness of the blankets in the tympan to one very thin piece of fine cloth; the lever has then very little tendency to return, and the pull is very easy in the extreme, requiring very little more force to move it at the latter, than at the first part: indeed, it is so different from the other press, that when an experienced pressman first tries it, he cannot feel any of the reaction which he has been accustomed to, and will not believe, till he sees the sheet, that he has produced any impression at all; and for many days after he begins to work at an iron press, he by habit throws back all the weight of his body in such a manner as to bring the handle up to its stop with a concussion that shakes his arm very much; and in consequence most pressmen, after a few hours' work, feel inclined to give up the iron press; but when they have got into a new habit of standing more upright, and applying only as much force as it requires, the labour of the pull becomes less than that of running the carriage in and out; and men who are accustomed to the iron presses only, would be scarcely able to go through the work of the old press.'

The management of *The Times* newspaper was quick to appreciate the value of the *Stanhope* press. While it could not reduce the need to set duplicate formes if the paper was to come out on time, it did speed up production by doing away with the two pulls for each sheet. What *The Times* itself described as a 'battalion' of *Stanhope* presses was used to print the paper for the first fourteen years of the nineteenth century.

The *Stanhope* shares with the *Columbian* press literary immortality—the *Columbian* in Arnold Bennett's novel *Clayhanger*, and the *Stanhope* in Honoré de Balzac's *Illusions perdues*. Jérôme-Nicholas Séchard, a master printer of the old school, in the first chapter, complains that the *Stanhope* is so rigid that it will batter the type. In fact, the *Stanhope* was used for 'fine' printing, producing a sharp impression of type and blocks. Hansard refers to the sharpness of impression produced by the *Stanhope* from formes of pearl and nonpareil letter (old small type sizes before the adoption of the point system). Charles Whittingham, the famous 'fine' printer, even went so far as to use the imprint 'The Stanhope Press' on a series of title-pages; and Thomas Bewick, the wood engraver, noted that the typography (in this sense the letterpress printing) of one of his books would be executed in the best style on one of the new *Stanhope* presses. There is no doubt that the *Stanhope* press was welcomed by those who wished to improve the standards of printing.

After a time the *Stanhope* press began to be superseded by the *Columbian*

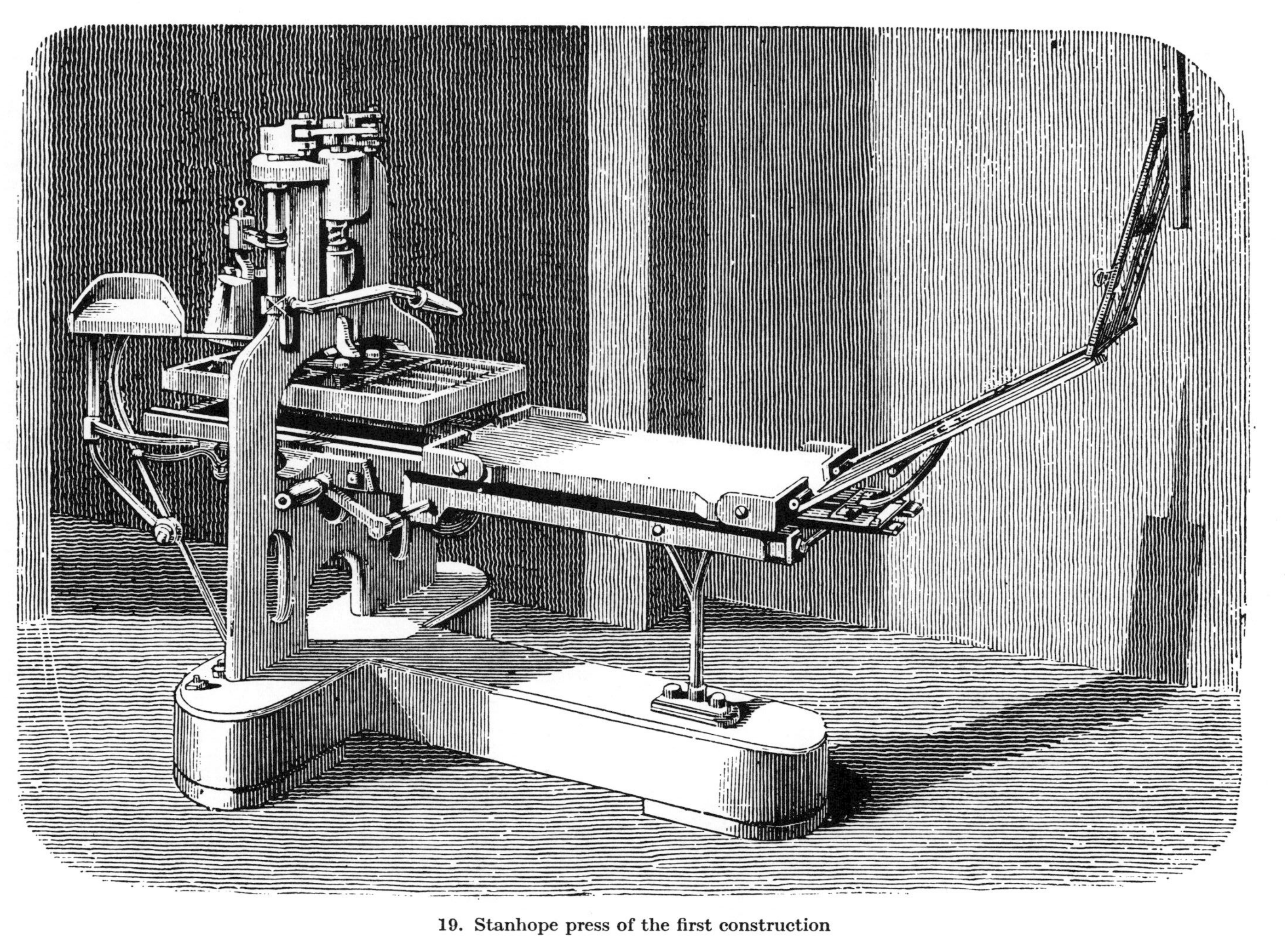

19. Stanhope press of the first construction

and *Albion*, although it appears to have been in production as late as the 1840s; and was still featured in catalogues twenty years later. The 1860 Harrild's catalogue advertises improved *Albion*, *Columbian*, *Imperial* and *Stanhope* presses. The illustrations are stock engravings and there is no suggestion that at that point Harrild's were actually manufacturing *Stanhope* presses. They had probably bought up the stock of an earlier manufacturer. By the 1880s the press was still in use in printing offices but had begun its relegation to the role of a proof press, a fate which also awaited its rivals. Nevertheless, in isolated instances, a *Stanhope* has been used as a production press up to the present day—a tribute to the solidity of its construction.

One man sought to out-Stanhope Stanhope in the use of power-multiplying levers. He was William Hope, an ironfounder of Jedburgh in the county of Roxburgh. He took out a patent in 1823 for additions and alterations to the press commonly known by the name of the 'Stanhope printing press'. He proposed the employment of an additional connecting rod and bent lever to that of *Stanhope* and thereby gain a proportionate increase of power in the press. It is interesting to note that the drawings accompanying the patent do not show the outlines of a typical *Stanhope* press, but one with a pillar surmounted by an ornamental urn at each side. The fate of Hope's super-*Stanhope* press is not known.

Lord Stanhope interested himself in many other fields of endeavour besides that of printing, to which, however, he paid a good deal of attention. His inventions in the composing room did not last, but from the point of view of the printing press his efforts to improve the inking of formes is worthy of consideration. A sheet headed 'Specimens of Typography, without the use of balls, executed at the Printing Press lately invented by Earl Stanhope. The Printing Press made by Mr. Robert Walker of Vine Street, Piccadilly. The Inking Roller made by Mr. Charles Fairbone, of New Street, Fetter Lane' bears the imprint of William Bulmer & Co, and is dated 1803. In the sheet, woodcuts appear inked by the new method, and, in a memorandum, Stanhope refers to 'the leather of the roller', which indicates that his rollers were not made from treacle and glue as they were later. He made experiments from all kinds of materials but did not achieve success. We know this from Hansard, who wrote: 'The late Earl Stanhope, when he invented the Printing-Press, which will bear his name to posterity, coupled with his object an idea of inking the forme on the press by means of a revolving cylinder; and in pursuit of this plan, spared no expense in endeavouring to find a substance with which to cover his rollers. He had the skins of every animal which he thought likely to answer his purpose, dressed by every possible process; and tried many other substances, as cloth, silk etc. without success. The necessary seam down the whole length of the roller was the first impediment; and the next the impossibility of keeping any skin or substance then known, always so soft and pliable as to receive the ink with an even coat, and communicate the same to the forme with the regularity required. All the presses of his early construction had, at each end of the table, a raised franch, type high, for the purpose of applying his rollers; but the obstacles interposed by nature herself totally baffled and defeated his lordship's plan in this respect.'

While Stanhope's experiments were not successful and hand presses continued to be inked with balls, the time was soon to come when rollers became a necessity for the new cylinder presses and Stanhope's ideas were revived.

The Stanhope press influenced others. The levers were used by George Clymer for his *Columbian* press (see Chapter 4), and Christian Dingler, of Zweibrücken, did not hesitate to use the characteristic T-shape wooden base from the Stanhope for his version of the Columbian, as the illustration below reveals.

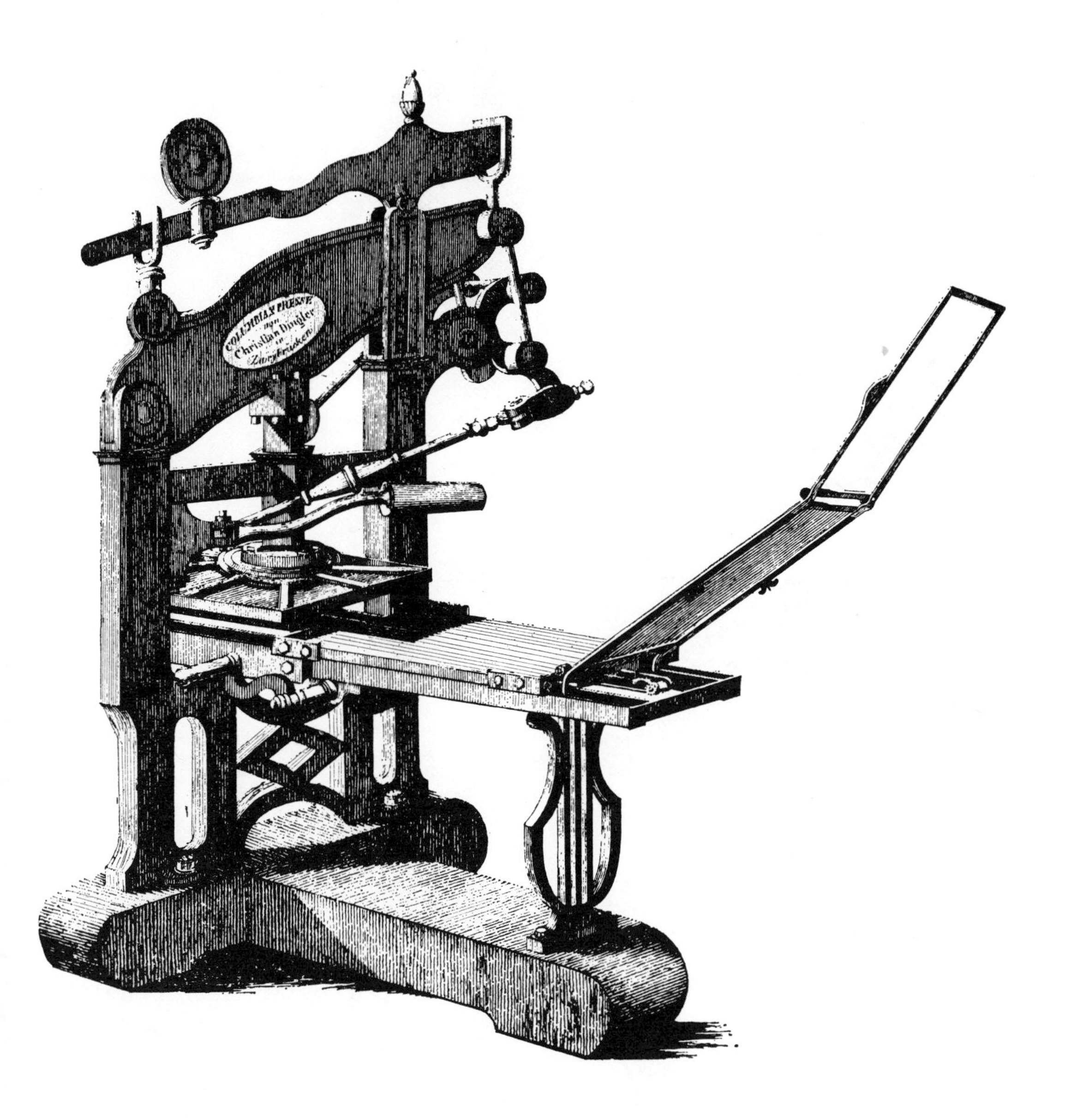

20. Dingler's 'Columbian' press with Stanhope base

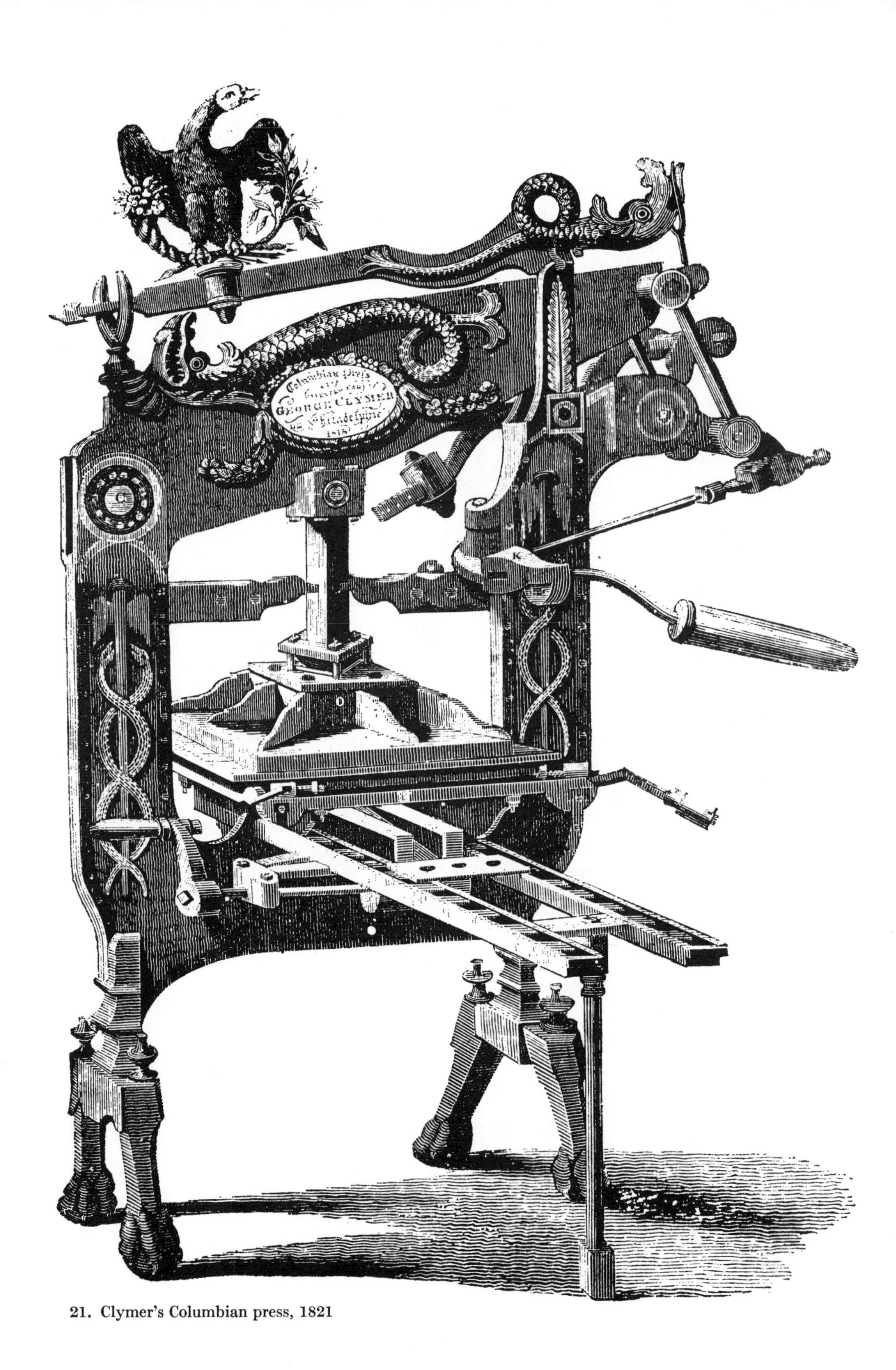

21. Clymer's Columbian press, 1821

4

THE COLUMBIAN PRESS

OF ALL THE nineteenth-century iron hand presses the *Columbian* is not only the most memorable by virtue of its striking appearance, but was the first to be manufactured in great numbers and by a wide variety of firms for a hundred years. It was invented by George Clymer of Philadelphia (1754–1834) who, taking advantage of the development in iron-casting techniques, chose to indulge in an orgy of symbolic decoration on his press.

This prompted the well-known passage in T. C. Hansard's *Typographia* (1825): 'If the merits of a machine were to be appreciated wholly by its ornamental appearance, certainly no other press could enter into competition with "The Columbian". No British-made machinery was ever so lavishly embellished. We have a somewhat highly-sounding title to begin with; and then, which way soever our eyes are turned, from head to foot, or foot to head, some extraordinary features present themselves—on each pillar of the staple a caduceus of the universal messenger, Hermes—alligators, and other draconic serpents, emblematize, on the levers, the power of wisdom—then, for the *balance of power* (we rude barbarians of the old world make mere cast-iron *lumps* serve to inforce our notions of the *balance of power*) we see, surmounting the Columbian press, the American eagle with extended wings, and grasping in his talons Jove's thunderbolts, combined with the olive-branch of Peace, and cornucopia of Plenty, all handsomely bronzed and gilt, *resisting and bearing down* ALL OTHER POWER!'

Hansard's description reinforces the suspicion that he never actually saw some of the presses he described. While the American eagle is shown in engravings grasping 'Jove's thunderbolts' these are absent from the talons in the earliest-known surviving presses. Whether or not they were present in the first American-built press, as an 1816 engraving might indicate, since no American press has been found, must be a matter for speculation. The fact that the bolts continued to be drawn by artists is simply an example of the familiar phenomenon of technical drawings being copied generation after generation without reference to the original object.

As Hansard points out, the caduceuses on the staple belonged to Hermes, the messenger of the gods, and represented the news-distributing function of the printing press. He was a little wide of the mark with his 'alligators, and other draconic serpents' as the creatures on the end of the counterbalance lever and at the top of the beam were supposed to be dolphins, which may have symbolized wisdom, since Apollo took the form of a dolphin when he founded the Oracle of Delphi; but the wise dolphins who helped fishermen are usually depicted as on

Aldus Manutius's pressmark and not as on Clymer's press. Clymer's look more like the 'dolphins' cast as handles on early cannon (and after which the cannon were named) which he may well have seen in old treatises when studying casting in iron.

On the beam was a cartouche of fruit and flowers to enclose the engraved name plate; but possibly the most distinctive characteristic was the American eagle grasping an olive branch and cornucopia, although no thunderbolts, which was so distinctive that generations of British pressmen, often unaware of the true name of the press, have referred to it as the 'eagle' press. Yet this very obvious American invention, in fact, made little headway in the land of its origin, achieving fame rather in Europe, and particularly in Britain.

The bizarre look of the press should, however, not be allowed to obscure the real technical and social advance which it represented. Technically, in part, it reverted to the earliest principle of the lever and fulcrum. This was combined with Lord Stanhope's idea of multiple levers to provide very powerful pressure on the platen with little exertion by the pressman.

Clymer was among a number of men in Philadelphia who began to make small improvements to the wooden printing press. He tried to change the method of cutting the screw, hanging the platen differently, and substituting rollers for cramps.

From a contemporary account in Abraham Rees's *The Cyclopaedia* (published in Philadelphia between 1810 and 1824) we learn that Clymer began these experiments in the year 1800, successively employing principles similar to those used by Medhurst and Roworth, but that his inclined plane press was succeeded in the year 1812–13 by the *Columbian* press.

Clymer's first ideas, therefore, were not original and it is not surprising that his claims to improvement were challenged by others. He decided to think the problem out anew and by 1812 had been inspired by the beam pumping engine, invented by Thomas Newcomen in England in 1712. Newcomen's overhead beam, known as the 'great lever', transmitted the motion of the piston to the pump, and Clymer adapted this 'great lever' (even using the same phrase in his patent) to a printing press to push the platen down by means of a square upright bolt or 'slider'.

The first Newcomen-type engine had been introduced into America in 1753 for use in the Schuyler copper mines in New Jersey, and Clymer was one of a group of Philadelphia engineers, chief of whom was Oliver Evans, claimed by Americans as the inventor of the high-pressure steam-engine. Evans obtained a copy of a book describing Newcomen's engine, so that Clymer could have known of Newcomen's 'great lever' even if he had not been acquainted with the engine in the Schuyler mines. In any case, he devised a pump of his own for clearing the cofferdams for the first permanent bridge across the Schuylkill in 1801. Indeed, he continued his interest in hydraulic engineering when he eventually went to England, where he patented his 'Columbian' ship's pump in 1818.

In the United States, the period between 1812 and 1814 had been devoted to experiments in relation to the *Columbian* press, which Clymer was trying to make efficient enough for newspaper printers. A letter from William W. Seaton, of Gales and Seaton, Washington, D.C., to William W. Worsley, of Lexington,

Kentucky, dated 20 March 1813, now preserved in the Draper MSS, State Historical Society of Wisconsin, explains that an iron press in the firm's possession 'the 2d or 3d that Clymer made', was superior in its greater strength and durability, the ease with which it worked and the uniformity of impression; but that it did not have a platen large enough to print a newspaper at one pull. The writer had been informed, however, that one used by the *Aurora* newspaper printed at one pull; and it is from the *Aurora*, published in Philadelphia, of 26 April 1814, that we learn that Clymer had completed two presses 'on different and distinct principles'. We know nothing of one but the other, the *Columbian*, became familiar enough. Clymer gave notice in April 1814 that his new iron press could be seen at William Fry's printing office, Prune Street, between Fifth and Sixth Streets, Philadelphia.

22. Pressmen at work on Columbian presses

The *Columbian* press bore little resemblance to anything which had gone before, either in technical details or in its highly decorated form. An engraving of 1816 shows a press with the nameplate on the beam surrounded by a rattlesnake. This is presumably what the American-built press looked like, but no actual specimen has survived for a comparison to be made. Clymer sold a number of presses to newspaper printers in New York and Philadelphia, one in Hartford and another in Albany, but that seems to have been the extent of the market, as his press did not meet with the general requirements of the time. The opening up of new territories in the United States demanded a light, easily transportable press, which the *Columbian* was not. The Ramage-built wooden presses were preferred by the pioneer printers. Having, therefore, saturated the eastern seaboard market, Clymer decided to see what Europe had to offer, and sailed for England in May 1817, at the age of 63.

Clymer settled in London and moved at a remarkable speed for those days, patenting his press in November 1817, and then looking round for a means of manufacture. According to a statement of Bigmore and Wyman in their *Bibliography of Printing* (London, 1880), Clymer entered into an engagement with R. W. Cope, of Lower White Cross Street, to make the press. This could not have prevailed because Cope began manufacturing his own *Albion* press in the early 1820s, and very early surviving *Columbian* presses bear Clymer's name alone, the earliest known being numbered 13, and dated 1818 (Plate XIV).

Clymer began to manufacture on his own by at least the end of 1818, when he rented a house and warehouse at No. 1 Finsbury Street, only a few minutes' walk away from Cope's premises. Clymer claimed that his press could be adapted to print the smallest cards or the heaviest formes by altering the length of the connecting rod by means of a screw which united it with the horizontal joint. He did not claim that the platen, sliding carriage, tympans, frisket, winch and rounce were his inventions. He could hardly have done this as each of these had been integral parts of the hand printing press from earliest days. But he did claim as his own 'the new arrangement and combination of those parts from which this press derives its superior effect'.

What did he mean by this? Presumably he was claiming that his combination of power-multiplying levers and great lever, acting on a fulcrum, was superior to other presses, and, in fact, as can still be tested, it did require less exertion on the part of the pressman, although the *Columbian* was slightly slower in operation than either the *Stanhope* or the *Albion*. The easier working was publicized by Clymer who, after working a *Stanhope* press, wrote: '. . . I can produce at the Columbian press, *properly adjusted*, a better impression from double demy or double royal forms, by a much lighter pull than is agreeable to a Pressman, than I can from any other Press I have tried.' A certain amount of bias is expected from the inventor, but among the testimonials published there is often reference to the fact that a boy of fifteen could work the press, James Moyes, of Greville Street, Hatton Garden, stating that the boy 'can pull the Press sufficiently to make a good impression from a heavy double royal form . . .'; with A. J. Valpy, of Tooke's Court, Chancery Lane, backing the judgement.

Six American printers had, in their testimonial, made the point clear: 'By an admirable arrangement and combination of mechanical power, the labour of press-work is astonishly reduced; and the well-founded complaints of the severity of that kind of employment are now totally removed by the introduction of the Columbian Presses.'

H. Hodson, Cross Street, Hatton Garden, referred to the 'amazing diminution of labour', while W. Marchant, of Ingram Court, Fenchurch Street, referred to a 'soft and easy pull'. *Kent's Weekly Dispatch,* of Sunday, 9 August 1818, echoed the sentiments of the American printers when it stated: 'The Pressman will no longer be subject to that *exhaustion* which has been the *opprobrium* of the trade . . .', and therefore Clymer could sincerely write: '. . . many excellent and experienced workmen, in the decline of life, may still continue to be employed at the Columbian press, in consequence of the ease with which they can execute their (hitherto laborious) work.' In this way Clymer was responsible for not only a technical but a social advance.

Clymer manufactured his presses in London in sizes between Super Royal and Double Royal. The first cost £100, the second £125 and any size in between £112 10*s*. In 1820 because of the 'extreme depression of business' he reduced these prices to £75, £95 and £85 respectively.

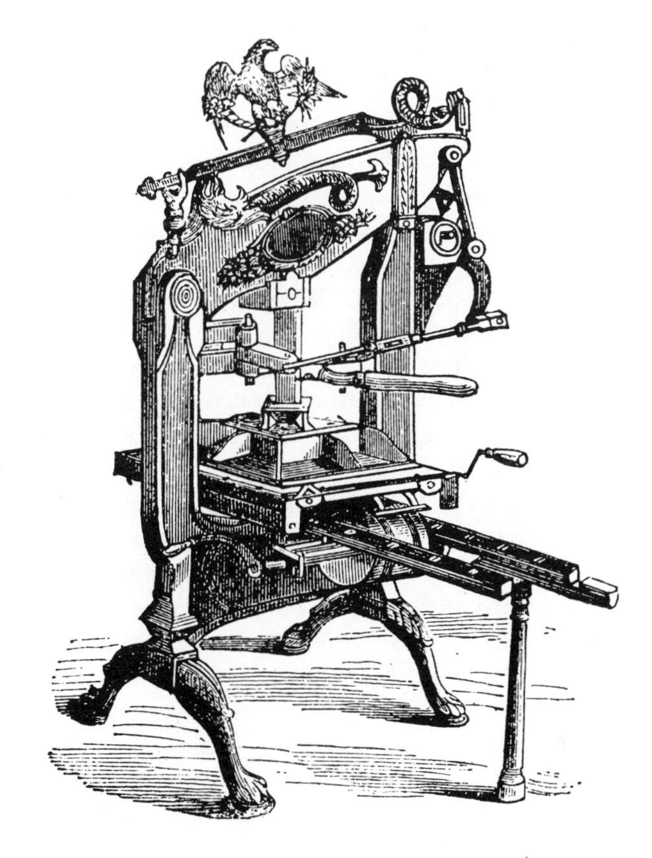

23. Typical stock block of Columbian press with blank nameplate, as used by distributors in their catalogues

Surviving presses indicate some of the sizes in which the *Columbian* was manufactured, but for other sizes there are only printed references. The smallest must have been the press which T. Cobb, of 19 Portugal Street, exhibited at the International Exhibition of 1851. This was so small that it would print nothing bigger than a visiting card. Charles T. Jacobi, the Chiswick Press printer, mentions a foolscap folio ($9\frac{3}{4} \times 15$ inches), and it is known that Richard Clay & Sons, Bungay, possessed one of this size in 1887. The largest size was probably the Extra Size Double Royal, advertised by Wood & Sharwoods, Ullmer, Harrild's and Lockett & Sons.

Clymer was a remarkable publicist. To begin with he made his product so distinctive that once seen it was not forgotten, and then followed this up with printed testimonials from various users. Those quoted are from a booklet, published on the appropriate date of the 4 July 1818. He made presentations of presses to reigning monarchs, a costly business, no doubt, but one which brought substantial rewards in the shape of 6,000 roubles from the Czar of Russia and a gold medal, valued at 100 gold ducats, from the King of the Netherlands. No reward is recorded from the King of Spain, but Clymer was able to use letters from the first two monarchs in further publicity.

The popularity of the press, resulting from its appearance, the publicity and its ease of handling attracted the attention of other manufacturers, but Clymer was protected for fourteen years by his patent, the expiry of which was waited anxiously by various entrepreneurs.

Clymer manufactured on his own at No. 1 Finsbury Street for nearly twelve years, taking a partner, Samuel Dixon, in the middle of 1830. By the end of the year the firm, now Clymer and Dixon, moved to smaller premises at 10 Finsbury Street. Clymer was now 76 years of age, and competition was growing not only from other hand presses but from the new cylinder press, and the patent had not long to go.

The inventor of the *Columbian* press died on 27 August 1834, Samuel Dixon carrying on the firm alone until 1845, when he brought in others to form Clymer, Dixon & Co. The firm stressed its priority in the field by using the words 'original patentees and manufacturers' on the nameplates of its presses. As a company the firm survived until Christmas 1851, during the year exhibiting *Columbian* press No. 1479, platen size 17×24 inches, at the International Exhibition. Competitors also showed their *Columbian* presses, including Henry Fairbank, of 121 Old Street, who said he would make them to order. At the end of 1851 the factory at 10 Finsbury Street was taken over by William Carpenter, who continued making presses until 1856. On the nameplates he described himself as 'late Clymer Dixon & Co', and used the 'Original patentees and manufacturers' formula. In his turn he was succeeded by Edward Bevan, who followed the same style of wording. Bevan kept the factory going until the end of 1862, and in July 1863, it having closed down, the sale of its contents took place. This event was announced in J. & R. M. Wood's *Typographic Advertiser* of 1 July 1863. Although it was not specifically stated, it seems that this firm bought presses and spare parts at the sale, and therefore advertised themselves as suppliers and repairers of the *Columbian* press.

But long before this happened Wood and Sharwoods had been waiting for the patent to expire, and after Clymer's death announced that they had installed modern machinery to manufacture the *Columbian* printing press, invented by the late George Clymer. Paying tribute to the potency of Clymer's name, Wood and Sharwoods not merely mentioned it in their publicity, but included 'Invented by the late George Clymer' on their nameplates, carefully omitting their own, but adding 'Manufactory 120 Aldersgate Street, London'. This has led to the mistaken idea that the 'Manufactory' was a continuation of Clymer's own works. Wood and Sharwoods even made a *Columbian* proofing press, with a small eagle perched on the beam, but this does not look as if it performed anything but a decorative function, although again it was good publicity.

24. Wood & Sharwoods' manufactory

Carpenter and Bevan were presumably old employees of Clymer and Dixon, who in turn took over the business. Another former employee, F. Mooney, set himself up at 8 Wellclose Street, St. George's East, and manufactured his own version of the *Columbian*. After a time many manufacturers and distributors of printing equipment began to issue *Columbian* presses. Thomas Matthews, a well-known distributor, of 78 Cow Cross Street, West Smithfield, actually described himself in advertisements as a 'manufacturer'. Others included J. C. Paul, James Cox and even Cope's successors Hopkinson and Cope. Typefounders, to the present day, have sold presses and other printers' necessities as well as type, and it is therefore quite understandable that such names as Reed and Fox, R. Besley (the Fann Street foundry), Vincent Figgins, the Caslon typefoundry and Miller and Richard should appear on the nameplates of *Columbians*. How many of these firms actually carried out the manufacture of the presses it is difficult to say. As in the case of the *Albion* press they may have shared the services of a foundry which cast the main parts. Lockett & Sons, who were press makers, may well have made their own *Columbians*, but there is no definite information on this. Naturally, printing supply houses would want to share in the profits to be made from a popular press, and at least four well-known houses, Frederick Ullmer, Harrild's, H. O. Strong and Thomas H. Hewitt, were responsible for distributing new *Columbian* presses well into the twentieth century, and Harrild's, in particular, with its characteristic version, may have had its own presses made at least as late as 1913.

The first *Columbian* presses in use in Scotland were those made by Clymer and Dixon, as the evidence of testimonials shows, but by the late 1850s Scottish engineering firms took up the manufacture, and D. & J. Greig, Ritchie & Sons and Thomas Long, engineer, all of Edinburgh, and John Ross, of Leith, are among the names found on surviving presses (Plate XV). Long and Ross appear to have joined forces. No attempt was made to alter the eagle counterweight, an American aspect which few could misinterpret, the Scots having no difficulty in retaining a symbol with such sales value.

In Europe, however, changes were sometimes made to suit national taste or the sentiment of the times. In France the eagle might not have been a popular symbol in the post-Napoleonic era, and was replaced in one case by a globe, and in another by a lion on a wreath, the cartouche on the beam in this case being dispensed with in favour of a figure of La Belle France. The manufacturers in Germany were not inhibited by Clymer's patent, and Fr. Vieweg & Son, of Brunswick, produced a version of the *Columbian* as early as 1824, replacing the eagle counterweight with one representing a griffin holding ink-balls, from the crest used by early German printers' guilds (Plate XVI). Christian Dingler, of Zweibrücken, who seems to have been an enterprising mechanic, produced a version with much of the decoration removed. The eagle counterweight was replaced by a simple disc. A feature of Dingler's press, however, was that it stood on a wooden 'tee', taken from the *Stanhope* press, and that the forestay was in the shape of a lyre, a conceit which Dingler used on other presses. A Dutch manufacturer, H. P. Hotz, of The Hague, in 1847 built a version in which the eagle is replaced by a lamp of knowledge, and the movable counterweight at the back bears an image of none other than the fabulous Koster, once the candidate for the role of inventor of printing (Plate XVII).

While Europeans modified the *Columbian*, London-made presses were exported to the mainland. Charles Manby Smith, in his autobiographical *Working Man's Way in the World* (London, 1853), describing a Parisian printing office of 1826, explains how he approached the pressroom overseer for a job, and writes: 'He led the way into a long room, extending, apparently, half the length of the street, where stood arranged in precise line a whole regiment of Columbian presses, of London manufacture, the number of which, as the crowning eagles rose and fell with rapid irregularity, I in vain essayed to count.'

The main change which took place in the construction of the *Columbian* press in its long career concerned the position of the bar. From about 1824 its pivot was shifted to the near side, as being more convenient for the pressman. The date is known from Hansard. Johnson, in his *Typographia* (1824), had complained that the bar was too far from the hand, and Hansard in his similarly named work, of a year later, also reported objections on this score; but added that he had just seen a press with the bar fixed to the near, instead of the off-side. The date is confirmed by a study of the oldest surviving *Columbian* presses. Some Scottish manufacturers introduced a cylindrical platen post, although there had been at least one French version of this earlier than 1860.

Clymer's original decorations, and the way in which his successors varied these, even slightly, can sometimes assist in the dating of presses which have lost their brass nameplates. On the oldest presses the great lever does not project at the fulcrum side; the flowers-and-fruit cartouche lifts up slightly on the left; and the legs are short and plain, ending in a claw and ball, but where they join the staple there are characteristic knobs or finials sticking up. These are easily broken off. Wood and Sharwoods seem to have copied this style, but Clymer's immediate successors, at 10 Finsbury Street, preferred 'cabriole' legs. Tradition dies hard, and the engraved brass plate continued to be part of the *Columbian* press until the late nineteenth century when the more utilitarian cast-iron plate, such as that used by J. C. Paul, became the fashion. Not until the Scottish manufacturers was there a great change in the decoration on the staple. Ritchie & Sons left the wings off the caduceuses, and substituted a square for a circle at the top of the right cheek. Their counterweight at the rear was in the shape of a shield. The Scottish manufacturers tended to perch the eagle on a slotted ball.

Thomas Long's great lever or beam is deeper than most, and the back bar, or stretcher, which is on all presses, is duplicated at the front, presumably for extra strength. The nameplate is in the centre of the bar. Flower patterns have been substituted for the caduceuses, and the legs have ball terminals. The secondary counterweight at the rear is cast in what is thought to be the head of Henri IV. So far, no explanation has been forthcoming as to its significance, if any. The small counterweight, in general, raises the question as to whether they were specially cast for *Columbian* presses or whether they were simply convenient castings found in a foundry at the time of manufacture. Early engravings of the press indicate a plain counterweight and it may be, therefore, that the distinctive 'Hope and anchor' design was not chosen by Clymer in America but was picked up by him in a London foundry.

Surviving presses, in fact, show that there must have been a trade in parts. While Long's presses are distinctive, as has been described, one survivor, bearing his name, has the eagle perched on a conventionally shaped base and the small counterweight is in the shape of 'Hope and anchor'.

Printed material indicates that the *Columbian* was manufactured well into the twentieth century. Ullmer's price lists of 1891, 1900 and 1902 show engravings of his 'Improved' *Columbian* 'with all the most modern improvements, and with improved connecting rod'. It is not possible to make out from the engraving exactly how the connecting rod differs from that on earlier models. The press, in general, looks much like others of the period, with leaf-pattern legs, terminating in paws. It ranged in size from Crown (21×16 inches) to Extra Size Double Royal (42×27 inches), and in cost from £26 to £75. Ullmer also showed 'the improved Columbian galley press', in two sizes—30×6 inches and 36×8 inches, priced at £12 12*s*. and £18. The counterweight was not in the shape of an eagle but resembled the cartouche of Ullmer's *Columbian* press, which carried the name of the firm. This galley press was probably the descendant of that made by Wood and Sharwoods. A similar press had been marketed since 1851 by Harrild's, without any particular name.

Harrild & Sons's *Catalogue of Machinery and Materials*, 1906, shows the final version of the *Columbian*. It has a massive base to the staple, unlike the original presses. The dolphin and cartouche have been retained, with a smoother-looking eagle bearing Harrild's trade mark on its breast, but otherwise the press is relatively undecorated. The legs have a leaf pattern and paw feet. Sizes ranged from quarto (10 × 7 inches) to Extra size (42 × 27 inches), and prices from £6 10*s*. to £80. By the time the 1913 catalogue had been issued the smaller sizes had been eliminated, and the range began at Foolscap broadside (19 × 14¼ inches) at a cost of £25. But a more significant change had already occurred by the 1911 catalogue when a difference in emphasis had taken place in nomenclature, indicating the changed role of the hand press. Both *Albions* and *Columbians* are listed under 'proof presses', and this role has continued to the present day.

It says much for the staying power of the *Columbian* press that more than 100 examples survive and are in use, some in institutions and schools, and in the workrooms of private printers, but a surprising number as proofing presses in commercial houses.

The *Columbian* press, retaining its major characteristics and its name, was manufactured by at least twenty-five British firms alone, but there were two other presses with their own patriotic-type names, which may have been inspired by the *Columbian*. The first was the *Britannia* press (Plate XVIII), although R. Porter, of Leeds, the manufacturer, could also have been influenced by the *Imperial* press. In his *Printer's Manual* of 1838, C. Timperley reports that the *Britannia* press was highly spoken of by many practical printers and extensively patronized in the counties of York and Lancaster. The date of the invention is not known exactly, but it must have been before Queen Victoria's accession to the throne in 1837 as the royal arms which form a decorative element incorporate those of Hanover. An example which has survived in the Science Museum, London, is dated 1835.

The iron frame of the *Britannia* is decorated on the front only. The cheeks are fluted square pillars surmounted by urns; the feet are in the shape of claws and the legs are splayed. A detachable lid bearing the royal arms may be lifted off the top of the press to gain access to the printing mechanism. A set of levers, resembling those on the *Columbian*, when put into operation, turn a cam in the head of the press to push down a piston, and hence the platen. The rear of the press is reinforced with an additional stretcher, and a counterweight to assist the return of the platen after printing is fixed to arms pivoted on the back of one of the legs.

The other similar press was the very strong Leggett's *Queen* press, the invention of an engineer in the employ of J. R. & A. Ransome, the agricultural machinery manufacturers, of Ipswich. Leggett put his ideas to his employers who, having satisfied themselves of the benefits of the press, and having the facilities available, proceeded to manufacture it.

The pressure was effected by a knuckle joint acting at the extremity of a very strong lever, which passed through mortices in the frame. The connecting rod from the bar handle to the knuckle joint also passed through the mortices and the power exerted at the centre of the press, with a gradually increasing intensity, produced the effect called by pressmen a 'soaking pull'. The most obvious borrowing from the *Columbian* was the counterweight mechanism, the counterweight itself being in the shape of a shield bearing the royal arms—a usual riposte to the American eagle. At first the Ransomes made one size only—28 × 21½ inches, at a price of £63, but later issued a Super Royal (29 × 21 inches) at £52 10*s*.; a Demy (24½ × 18 inches) at £42 and a folio foolscap (15¼ × 10¼ inches) at £16 16*s*.

As with the *Britannia* it is not possible to give an exact date for the invention. A *Queen* press made by 'Ransomes and May' was shown at the Great International Exhibition of 1851, and a model bearing the same name was shown on the stand of Cowell's, the printers; but a surviving press carries the names of J. R. & A. Ransome (as do two advertising leaflets in the Ransome archives). Since the firm took the name of Ransomes and May in 1846 (changing to Ransomes and Sims in 1852) the press was probably first built in the 1840s.

In 1827, when he was 73 years of age, Clymer patented another press, described as 'an improvement in the typographical printing between plain or flat surfaces'. It was an endeavour to penetrate the newspaper printing field, which, by then, was gradually becoming used to the cylinder press. Clymer stated in his patent: '. . . by this new-contrived press, I propose to print two forms of double royal paper at one time, being a surface of four feet six inches by three feet three inches, which is twice the size of the largest newspaper at present printed'.

As far as can be gathered from the specification, Clymer was influenced by other inventions, such as the *Albion* press, which utilized a toggle joint, although there are elements of his former ideas, such as the guided piston. A massive toggle joint was attached to the side of the press by two iron bars, and when the handle was swung outwards it brought the two parts of the joint into a straight line, forcing down the piston on to the platen to make an impression. When the handle was moved back the platen rose but, in order to assist the motion, Clymer proposed suspending it by rods to weighted levers at the top of the press 'which balance the platen, and enable it to be worked with very little exertion of the pressmen'.

He also adopted an idea from John Brown's patent of 1809 by proposing two tympan carriages, one on each side of the press, and a stationary forme under the platen. The idea was to ink the forme by an 'elastic inking roller', introduced at the side of the press, and while this was being done, sheets of paper were to be placed on the tympans, one of which was to be run in on side rails above the forme. The platen was to be brought down, and an impression taken, the platen raised, the forme inked again and the other tympan run in for the performance to be repeated.

One point is noticeable. Clymer obviously felt it no longer necessary to emphasize an American origin. The press was to be unadorned except for claw feet. The weights on the platen levers resembled ordinary printers' paper-weights. No record has been found which indicates that this press was built or marketed. A similar, if more advanced, approach was made on the Continent with the *Selligue* press (see Chapter 5), but Clymer's press was too complicated for what it was intended to achieve, and would hence be wasteful of labour. Those British newspaper proprietors who were interested in speed of production would, at this point, be turning to the cylinder press.

Clymer's presses provide an object lesson. The 1827 press lacked all the attributes of the *Columbian* and sank into obscurity, whereas the *Columbian* carried on for more than a hundred years.

25. This 1836 illustration shows the 'acorn-shaped' frame associated with early American iron hand presses. (See page 76)

5

THE IRON HAND PRESS AFTER STANHOPE AND CLYMER

FOLLOWING THE TWO pioneers, Lord Stanhope and George Clymer, engineers on both sides of the Atlantic, from 1819 onwards, began to produce iron hand presses in large numbers. In Europe manufacturers tended for a time to copy British and American presses.

While inventors rarely departed from the essential characteristics of the traditional printing press, without exception they abandoned the screw as a means of providing the downward pressure on the platen. Whatever device they chose, however, they were faced with the same problem as Lord Stanhope and were obliged to consider the form of the main frame or staple, as it was vital that this should be strong enough to withstand the increased power applied. There was a conflict between this requirement and that of ease of transport, and a variety of solutions were propounded.

The all-metal presses which emerged were no faster in operation than their wooden predecessors, although they were capable of producing better-quality work. That is not to say that Brown's notions (see Chapter 2) of a faster rate of printing with two tympans or with automatic inking, were completely abandoned, but, in the main, the concentration was on the printing mechanism. Greater speeds came only with the cylinder press, and automatic inking on hand presses tended to be more trouble than it was worth.

The success or otherwise of a press depended on more than its efficient functioning. There was the familiar need for financial support, manufacturing capacity, publicity and marketing. Among the presses which may have seemed, at the time of their invention, to have a future were the *Ruthven* and the *Cogger*, but they did not, apparently, achieve sufficient popularity to be among those which continued to be manufactured to the end of the nineteenth century.

As early as 1813 John Ruthven, an Edinburgh printer, patented an iron press in which the bed carrying the type remained stationary while the platen was moved over it on a wheeled carriage. Springs kept the platen raised until the moment of printing, when the power was applied through a series of levers worked by depressing a bar at the side of the press. Ruthven's fellow Scot, Adam Ramage, who must have kept in touch with his homeland, was sufficiently impressed with the *Ruthven* to obtain the rights to manufacture it in the United States, but he seems to have abandoned it in favour of the more popular type of iron hand press when these began to appear.

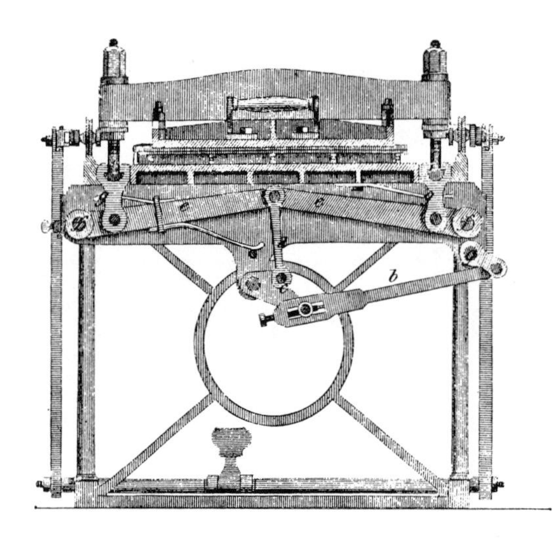

26. Ruthven press

The invention of T. Cogger, of Wardrobe Terrace, Doctors Commons, London (not far from *The Times* office), was an iron press with wrought-iron cheeks or perpendicular pillars, a low head and a compound lever beneath it (Plate XIX). The compound lever drew its end about one-quarter round a collar into which were fitted two studs of case-hardened iron with convex faces which, when put into motion by the lever, moved up inclined planes of different angles of inclination, so that when the platen was first put into the descent, and power was less needed than velocity, the latter was rapid; but as the platen arrived towards the point of pressure the velocity was diminished to increase the power. Hansard devoted nearly a page and a half to the *Cogger*, and a large illustration was included in his *Typographia*, but the press does not appear to have been a success. The studs on the inclined planes would have been subject to heavy wear, and this may be the reason why inventors thereafter tended to avoid the Roworth-de Heine-Cogger (see Chapter 2) printing technique in favour either of Medhurst's ideas or, to the greatest extent, the knee joint.

A German press resembling the *Cogger*, but relying on two ball counterweights rather than springs, was made by C. Hoffman, of Leipzig, from 1826 onwards and between 1832 and 1833, Friederich Koch, of Munich, copied the *Cogger* calling his press the *Säulenpresse* (pillar press).

While engineers and mechanics were seeking markets for their presses, individual nineteenth-century printers, both professional and amateur, invented presses for their own use. In most cases there are only printed references to guide us as to what these were like. Peter Buchan (1790–1854), collector of Scottish ballads, according to the *Dictionary of National Biography*, constructed in 1819 a new press on an original plan. It was to be worked by the feet instead of the hands, and could print from stone, copper and wood as well as from type. No further information had been obtained about this press, which could have been a forerunner of the jobbing platen, with its treadle action. The *D.N.B.* article continues: 'Buchan also invented an index-machine showing the number of sheets worked off by the press, but an Edinburgh press-maker borrowed this invention, and, taking it to America, never returned it to the inventor.' This would have been an early device for calculating the number of printed sheets. Joseph Bramah demonstrated the possibility of printing numbers on sheets as early as 1806, and others followed, but, subject to further research, an index to show the number of sheets printed did not become a practical proposition until the development of the cylinder printing machine. The *Typographical Gazette* recorded in June 1846: 'Mr. Edward Strong has contrived a printing machine index to register every copy to the extent of 100,000', and a registration index was patented by Robert Clegg in 1857.

Also, early in the nineteenth century, the Revd. William Davy, curate of Lustleigh, in Devon, printed a number of books on a press of his own construction, about which little is known. In 1930 there remained only some relics, comprising pulley wheels and rope, and a number of rollers, which may have been parts of his press.

William Glazier, of March, Isle of Ely, wrote to the *Expositor*, a weekly journal, on 14 March 1851, to say that he had recently contrived a press in which the plan was 'to run the bed beneath the platten by turning a handle with the left hand, and then pulling another handle with the right to give the impression'. Immediately the tympan was turned down on the forme a single pull of the handle brought the platen down with an equal and powerful pressure, and when the handle returned to rest the tympan could be thrown up, 'and thus a considerable number of impressions beyond any other hand-press might be taken'. Mr. Glazier had also, according to his letter, invented a method of printing two colours at once, and was seeking capital support. Presumably he was unlucky as his inventions do not seem to have reached the stage of general use.

However, at least one example of a 'home-made' press still survives in the Longford-Westmeath County Library, Mullingar, Ireland, thought to have been made by John Lyons, a printer who was also a self-taught mechanic and clock-maker (Plate XX). For more than a century the press remained at Ledeston House, when in 1964 it was removed to the County Library. The wooden table was badly worm-eaten and has been replaced by an angle iron base, while the uprights which carried the tracks for the bed for the press were welded to the new base top. The name 'J. Lyons' is stamped on the spindle, and there are four brass ornaments on the upper side of the platen. When the then Librarian, S. Ó. Conchubhair, received the press there was a piece missing which had actually conveyed the pressure from the lever on to the spindle, and he had a piece specially cast as a replacement.

The bed of the press is richly ornamented at either end with a leaf-type ornament. A number of books were printed on the press between 1827 and 1853, and the Library has copies of most of them.

The lever at the top of the press moves on a shaft between two uprights, and pressure can be affected in either direction—forward or backward. Basically the movement is of the lever and fulcrum type, notches in the lever fitting into a slot of the element which pushes down the spindle and hence the platen. Printing must have been rather slow and tiring.

As the mechanical skill of inventors often exceeded their financial standing, many presses may have suffered the fate of Mr. Glazier's, but some were saved from oblivion by a farsighted manufacturer, and the career of Richard Hoe, of New York, is of significance in this respect, since many presses associated with his firm in the nineteenth century were by no means his inventions. He was acknowledged by contemporaries to be an ingenious engineer and a good businessman, but there was a feeling, perhaps unjustified, that he exploited the work of others rather than inventing anything himself. Various presses which started off as independent productions ended up in Hoe's catalogues, but Hoe, naturally, claimed that they had been improved.

It is sometimes difficult to allocate priority to a particular invention since an idea or specification has to be translated into a working device, and a successful one at that. Complaints of 'stealing' have already been mentioned in this narrative and will continue to occur, but often the complainants did not seem to have realized that a particular technique may be common knowledge—indeed may have been for hundreds of years.

Often, as with Lord Stanhope's press, the novelty arose from applying some well-known system—in his case compound levers—to an existing device—a screw press. George Clymer adapted the compound levers to an even older conception—the beam working on a fulcrum. This did not stop critics implying that the presses had been copied from others. Because Stanhope's iron frame resembled that of the *Haas* press, it was said his press was a copy; and Hansard, in discussing Clymer's *Columbian*, wrote: 'Whether it be entirely original, as to principle, is a matter now to be considered. Some years ago a Mr Moore invented and took out a patent for a press, the power of which was gained by the fulcrum and the lever, instead of the inclined plane or screw. I had one on a small scale (foolscap size) which was made by Mr Arding for the late Mr Rickaby some years after the expiration of Mr Moore's patent, the power of which was gained by a wheel and chain, worked by the left hand, which drew down a strong lever projecting through the cheek of the press, having its fulcrum moveable (the pivot being fixed) upon the centre of the platten. The Columbian press I conceive to be upon the same principle; the lever being brought down upon the fulcrum (the platten) by a combination of levers, with the right hand, instead of being brought down by the wheel and chain.' Since it was up to anybody to use a known method of applying power, Clymer would not have denied that his press was based on the fulcrum and lever, but his use of compound levers to bring down the giant lever was novel and far more powerful than Moore's simple lever or Arding's wheel and chain.

The question of originality or otherwise again arises when considering the application of the toggle joint to the hand printing press. A mechanical toggle joint is a device consisting of two bars or levers joined together end to end, but not in line, so that when a force is applied to the knee join the tendency is to straighten the bars, the parts abutting or jointed to the end experiencing endwise pressure. If the top bar is fixed, increased pressure will be applied by the lower bar—a useful method of pushing down the platen of a printing press.

A version of the device, the 'torsion toggle' has already been mentioned in relation to the *Medhurst* press, and the *Albion* press, invented by R. W. Cope in about 1820, which will be dealt with separately, utilized a characteristic toggle, probably developed independently of those in the United States, where the device achieved most prominence. In any case, the printing-press makers did not invent this mechanism but adapted it from simpler domestic presses.

John Wells, of Hartford, Connecticut, had adapted the toggle or knee-lever action to a wooden press as early as 1816. He was an inkmaker and it is interesting to note that he obtained his idea from a press possessing a lever and simple toggle joint in his linseed-oil factory. Like others, he discovered that the increased power was too great for a wooden frame to withstand and, accordingly, he patented an iron toggle-joint press in 1819, incorporating long toggle levers, claiming that their length gave unusual power. They were pulled into a straight line by use of a bar. This press had an iron frame with parallel sides and, originally, a heavy iron ball acted as a counterweight to return the platen, but Wells substituted springs for this purpose later. In the same way as Clymer he at first located the bar on the offside, making printing difficult, but then corrected his mistake by moving it to the near-side. At least five Wells presses survive, and the one in the Smithsonian Institution, Washington, D.C., is particularly interesting as it has a ball counter-weight and the bar is on the offside (Plate XXI).

The *Wells* press was on the market for only a few years when two rivals appeared —the *Smith* and the *Rust.* Adams's *Typographia* (1828) carries a complaint by Wells that a Peter Smith had copied his press. Whether this was so or not, it is true that the partnership of Robert Hoe and his brothers-in-law, the Smiths (formed in 1805), put the *Smith* press on the market in 1821 not having patented a press until the December. At nearly every stage in the development of the printing press and machine the firm of Hoe & Co. was involved in controversy, and at this distance of time away it is not always possible to distinguish the true facts of a situation. However, we are fortunate in that Stephen Tucker, apprenticed to the firm in 1834 (the year after the death of the founder, Robert Hoe, senior), decided to write a firm's history (which exists only in manuscript form). From this, factual statements on Hoe's activities can be obtained, and his version of the Hoe–Rust relationship following the Hoe–Smith dispute will be given. But first there should be a consideration of what is thought to be the characteristic 'acorn shape' of American iron hand presses.

There seem to have been some British-American exchanges in this period of printing-press development, about which little has been recorded, although a certain amount can be conjectured. Such figures as Ramage, Treadwell and Clymer clearly involved themselves on both sides of the Atlantic but Clymer, in particular, is also thought to have used the close ties which the Quaker community of London had with that of Philadelphia, then the premier city of the United States.

The Quaker family of Barclay, bankers and merchants in both London and Philadelphia, seems to have been involved with printing equipment. In the 1790s George Barclay & Co., of London, acted as agent for Mathew Carey, printer of Philadelphia, and exported equipment to the United States for him. It is consequently not surprising that a merchant, David Barclay, 'of Broad Street' (presumably of the firm of Barclay Brothers, 34 Old Broad Street, City of London), should take out a patent for a printing press in 1821. In the patent Barclay refers to the invention of 'a spiral lever, or rotary standard press' communicated to him 'by a certain foreigner residing abroad'. At this point in time the invention is unlikely to have been from any other country but the United States, and, in any case, the *Encyclopaedia Londinensis* (1825) specifically refers to *Barclay's American press*; and it is quite possible, because of his Quaker connections, that Clymer was the 'certain foreigner' and that the press was one of his earlier inventions. Details could have been sent to London for patenting because it was too similar to other American inventions. Naturally, the press could have been that of Fairlamb, but this was a wooden structure and Barclay's was clearly of metal, with a very distinctive 'acorn-shaped' frame. This type of frame has always been associated with early American hand presses, being first used by Peter Smith in the same year as Barclay's patent. This may be a case of two men, widely apart, looking for a strong frame and coming to the same conclusion—that something like Lord Stanhope's second bow-shaped staple was needed. Since there was no need for space for a screw at the top, the frame continued until it joined up making the acorn shape almost accidentally, Smith's, with its indentations, being more pronounced than Barclay's.

The Barclay press was based on the principle of inclined planes—the fanciful description of 'rotary standard press' being slightly misleading as in essence it was simply a platen press for which Barclay did not claim privilege except for the mode of obtaining pressure.

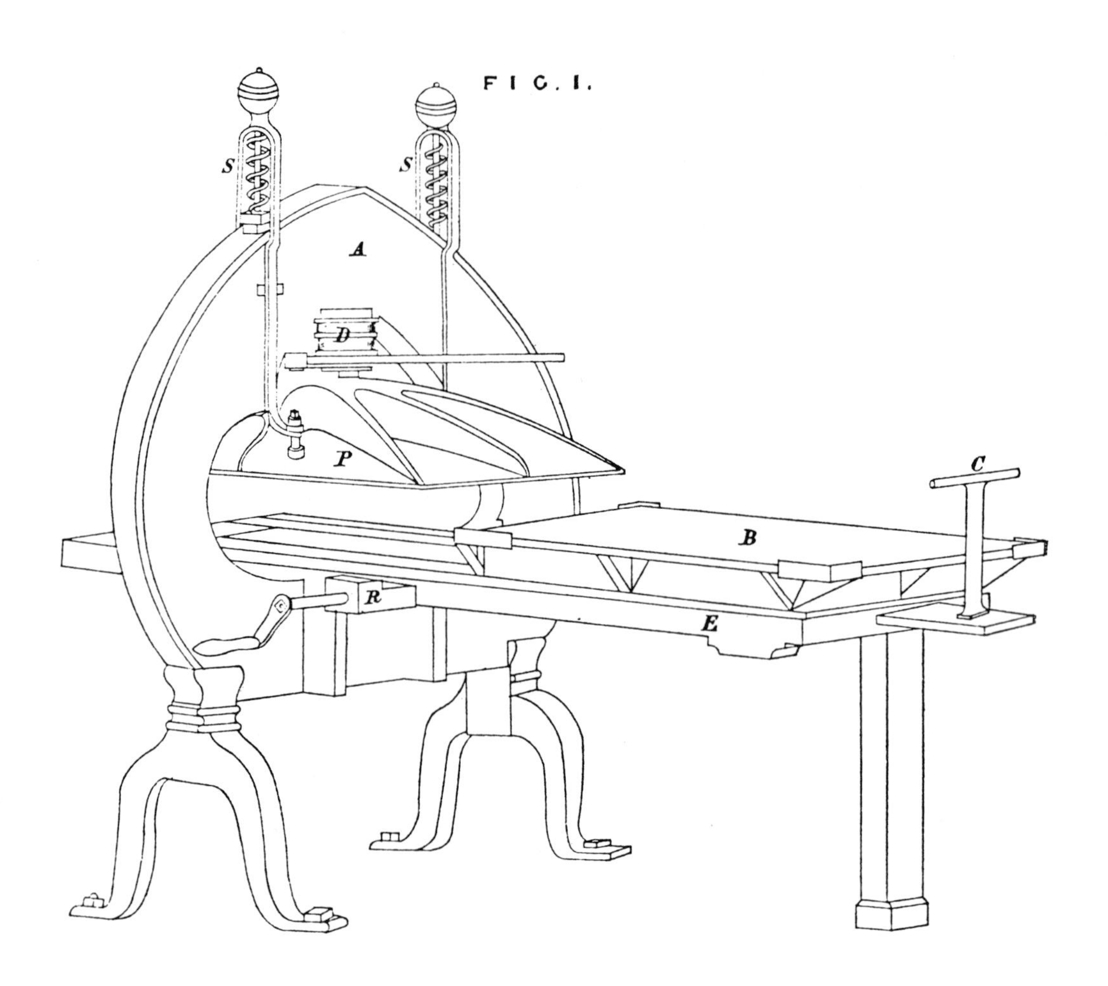

The mechanism used consisted of inclined planes and rollers, which Barclay called 'the rotary standard'. One inclined plane was attached to the top of the movable platen and another to the top of the frame of the press, both having two grooves in which the rollers rested. Between the rollers was a wedge similarly notched at each side for the rollers. Attached to the wedge was a protruding arm fixed to a series of levers and a bar. The effect of pulling the bar was to make the rollers rise out of their grooves and pass along the inclined planes, while the thickest part of the wedge, coming under the rollers at the same time, the platen became depressed to make room for it. The platen was returned after each printing operation by spiral springs.

The *Smith* press had the distinctive heavy cast-iron frame, and was operated by a toggle joint with equal-length levers, a characteristic being that the knee joint was drawn in, whereas in other presses it was pressed in (Plate XXIII*b*). The press was said to have the disadvantage that if the bar were let go suddenly it would fly back with so much force as to cause its parts to jump from their sockets.

The toggle joint action and the *Smith* acorn-frame attracted other manufacturers. Among these were three separate Bostonians, Phineas Dow, Otis Tufts and Seth Adams, each of whom produced a hand press with equal-length toggle levers and a heavy iron frame in 1827, 1830 and 1831 respectively. Dow's press, known after its inventor as the *Couillard*, differed slightly as it had a more oval frame than the others. Seth Adams's press No. 248, bearing the name 'Seth Adams & Co, Boston', survives in the office of the *Vineyard Gazette*, Edgartown, Massachusetts. Seth was joined by his brother, Isaac, in 1836, and press No. 329 with the inscription 'I and S. Adams' is in the Special Collections department of the Providence Public Library, Rhode Island. Press No. 335, in the same library, carries the inscription 'I. Adams & Co, Boston'. Another Bostonian, Samuel Orcutt, in 1840 used the acorn frame as the basis of a small card press.

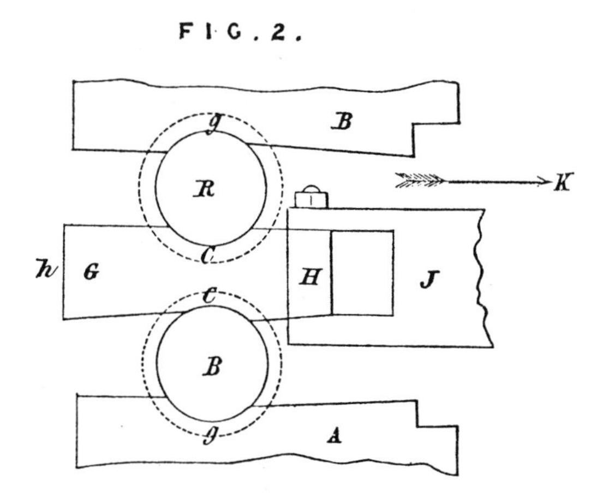

27. (*opposite*) Diagram of Barclay's press with its distinctive frame (A)

28. Diagram of pressing mechanism on Barclay's press. When the bar (J) attached at (H) was pulled in direction (K) the rollers (R) and (B) left their grooves (gg and CC) and moved along the inclined planes A and B. When the thickest part of the wedge (G) came between the rollers it forced down the inclined plane (A) and hence the platen to which it was fixed

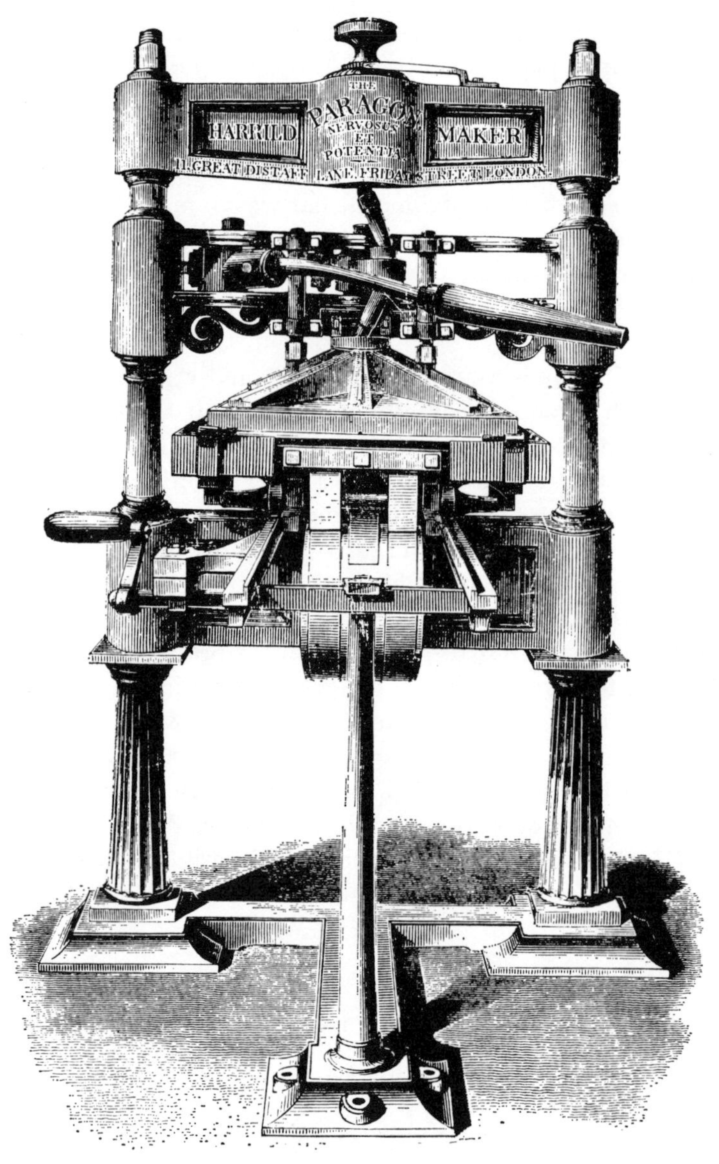

29. Harrild's Paragon press. Engraving by his son-in-law, George Baxter

The one British press which seems to have incorporated the 'American' type of toggle motion is the *Paragon*, made by Robert Harrild, then of Great Distaff Lane, London. Little is known of the press except that Harrild, a printer, changed his business to that of printers' dealer in 1832, selling an assortment of presses, among them his own *Paragon*, which was said to be built 'upon a particularly simple and novel construction'. Since it was thirty years before Harrild seriously entered into the manufacture rather than the mere supply of presses, the *Paragon* may have been a preliminary experimental venture on his part.

One iron hand press which has defied identification, despite an unusual characteristic—the bed is moved under the platen by lever action—is the so-called *Moffat* or *Kuruman* press, now in the Kimberley Public Library and considered to be the oldest iron hand press in South Africa. The press carries no maker's name or identification, but it is known to have been sent to South Africa by the London Missionary Society in 1825. It was given by Dr. John Philip to the missionary Robert Moffat who transported it from Port Elizabeth to Kuruman in 1831 (Plate XXIII*a*).

The frame is of the rounded kind associated with early American presses, but which, as had been mentioned in relation to the Barclay press, may have been derived from the *Stanhope*. Pressure is exerted by lever and toggle joint and the platen is lifted by a horizontal leaf spring at the base of the frame acting through levers. The bed is not moved on rails, being mounted directly on to a heavy lever. When the handle is turned the lever propels the bed under the platen. Directly beneath the platen and bolted to the frame are two U-shaped solid iron blocks on which the bed rests during the operation of printing.

This press has been investigated a number of times and inquiries have been made over more than half a century as to its origins, but with negative results. No definite statement can be made as to whether it is of British or American manufacture, but the absence of any maker's name or mark leads to the presumption that it was an experimental effort, incorporating characteristics of other iron hand presses of the time, but utilizing an original bed movement.

There is no reason to suppose that other inventors did not try to emulate Stanhope, Clymer, Wells, Rust or Cope, but physical proof of their work rarely survives. The *Moffat* press has done so, but unless new information is discovered, its maker's name will remain a mystery.

A key figure in the development of the American hand press, Samuel Rust, of New York, came to the conclusion that the heavy frame was a drawback in transporting a press, and devised a light-weight frame which could be taken apart. However, when, in 1821, he patented his *Washington* press (which was to become as famous as the *Columbian* and the *Albion*), he used an acorn-shaped frame. The toggle differed from that of the *Wells*—that is the levers were not of equal length and were joined at the top in a triangular shape. The bar for pulling which crossed them made a figure 4, and the toggle is so described, as can be understood from an examination of a *Washington* press. After 1829 Rust built his press with a straight-sided frame, but instead of it being all cast iron it had the uprights at the side hollowed for admission of wrought-iron bars which were riveted at top and bottom of the casting to provide increased strength and to reduce the amount of metal required. It was claimed that the *Rust* frame was half the weight of the *Smith* 'acorn'.

Hoe's had their eye on Rust's *Washington*, which was manufactured by the firm of Rust and Turney. Rust refused to sell out to them and so in 1835 (according to Tucker) John Colby, foreman of the Hoe hand press and jobbing rooms, pretending to set up in business for himself, succeeded in buying Rust's patent right, stock, tools and shop complete. He continued the manufacture of the presses, but shortly transferred the whole business to Hoe's. Hoe thereupon began to manufacture the *Washington* press, the platen and frame becoming heavier as time went on. It was made in seven sizes and began to outdistance its rivals, even a Hoe-built 'Improved Washington'. The *Smith* press continued to be made by Hoe, but

Patent Washington Printing Press.

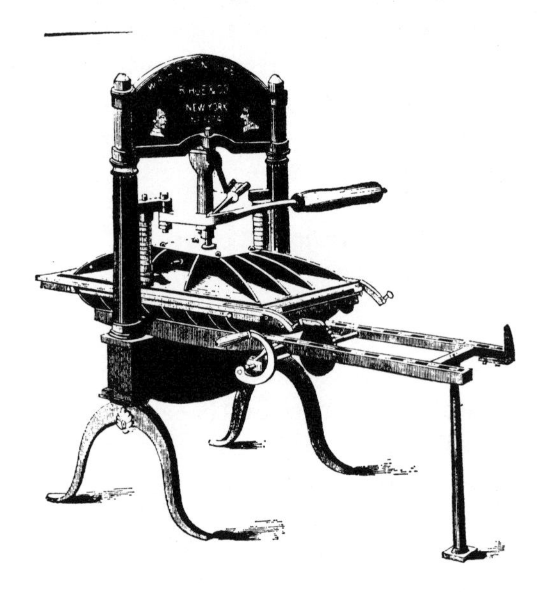

THE celebrity which our Patent Washington and Smith Hand-Presses have obtained during the last forty years, renders any remarks upon their superiority unnecessary. They are elegant in appearance, simple, quick, and powerful in operation, and combine every facility for the production of superior printing. Each press is tried at the manufactory, and warranted for one year.

DIMENSIONS AND PRICES.

	Bed.		Matter.		
No. 1,	17 x 21	inches,	13 x 17	inches,	$
No. 2,	20 x 25	"	15 x 20	"	
No. 3,	24 x 29	"	19 x 25	"	
No. 4,	26 x 34	"	21 x 29	"	
No. 5,	29 x 42	"	24 x 37	"	
No. 6,	32 x 47	"	27 x 42	"	
No. 7,	35 x 51	"	30 x 46	"	

Price includes two pairs of Points, one Screw Wrench and Brayer, one Slice, and one extra Frisket.
If the frame is made to be taken to pieces, $ extra.

Grand, Broome, Sheriff, Columbia, and Gold Sts., New-York; and Dorset St., Salisbury Square, London, England.

30. Early announcement for Washington press, which carried medallions of George Washington and Benjamin Franklin

was eventually discontinued. The *Washington* was so popular that other makers began to copy it, but they are too numerous to mention in detail. The *Reliance* group of presses, made by Paul Shniedewend, of Chicago, are dealt with in Appendix II (The proof press). The *Perfection* proof press, *Washington* style, was manufactured by H. B. Rouse & Co., Chicago, as late as 1926.

The Hoe 'Improved' *Washington* was launched in 1857. This press, with the toggle beneath the bed to push it upward against the platen, had an erratic history. The original idea was that of James Maxwell, of New York, who made his *Eagle* press on this basis from 1836. It is doubtful whether many were sold, and the idea was revived by Guilford and Jones, of Cincinnati in 1851. The next year Charles Foster, of the same town, obtained a patent covering the same kind of arrangement. Then Foster was bought out by Hoe in 1857, and the patent was used to design the 'Improved' *Washington*, which was supposed to supplant the Rust-based *Washington*. This did not happen, and the 'Improved' version was discontinued in 1868.

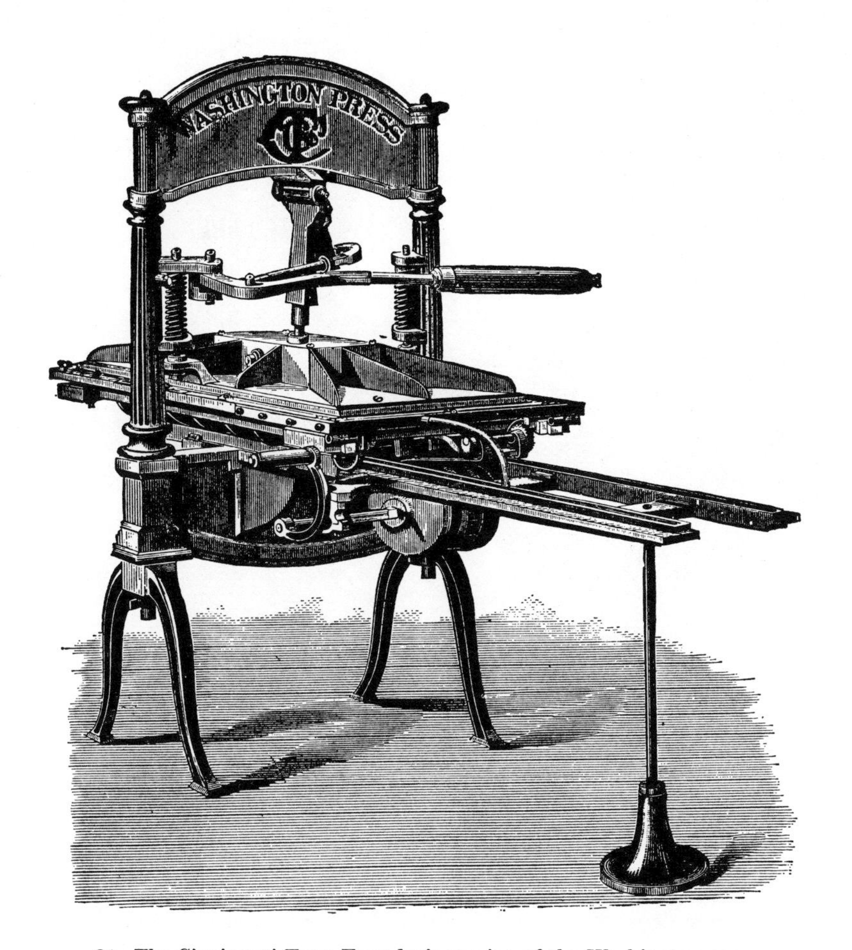

31. The Cincinnati Type Foundry's version of the Washington press

32. Unidentified hand press illustrated in a Russian book (1845–50), unlikely to be of Russian manufacture. It appears to be a screw press with 'Stanhopean' levers and a fly at the top of the screw

33. (*opposite*) Dingler's Zweibrücken press.
Compare decorative elements with his Hagar press. (See page 86)

A man who should not be overlooked in the development of the American iron hand press is Adam Ramage, although his greatest achievement belongs to the period of the wooden press. By the 1830s he was feeling the competition of the iron press and proceeded in 1833 to construct one of his own—the *Philadelphia,* which had a distinctive wrought-iron frame with a triangular top. The power was provided by equal-length levers with extensions. He also made, in about 1845, a press invented by Sheldon Graves, called the *American.* This was of the *Washington* type. Ramage's successor in business, Frederick Bronstrup, continued to make the *Philadelphia,* under the name of the *Bronstrup* press from 1850 to 1875.

In Europe the toggle joint was copied—a typical press being the *Zweibrücken,* named after the town in which the manufacturer, Christian Dingler, worked. Dingler was among the most active of German hand-press makers. Friedrich Koch, of Munich, called his production the *Kniehebel* (knee lever) press.

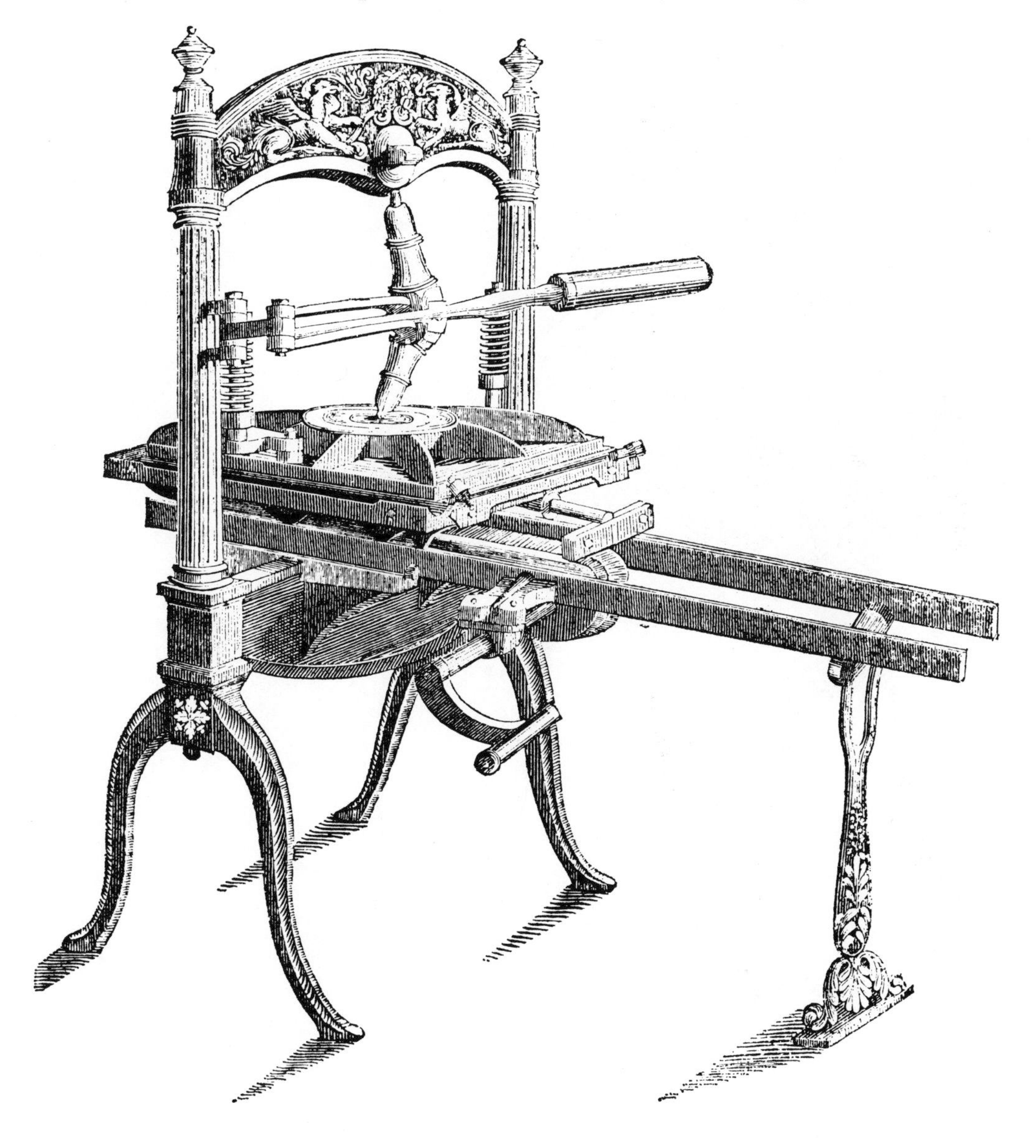

Another direction was taken by Abraham O. Stansbury, a New York bookseller, who became interested in printing. John Wells, who, as already mentioned, complained about Smith in Adams's *Typographia*, stated in the same publication that in 1819 Stansbury had examined his experiments. In the event, Stansbury did not follow Wells but adopted Medhurst's idea of the torsion toggle (see Chapter 2). Whether he knew of Medhurst's mechanism is not known and, indeed, in general, Stansbury is a shadowy figure. From surviving presses it can be conjectured that he constructed a few presses in New York, which differed little from the old wooden presses except for the toggle mechanism; and that he then substituted cast iron for the bed and platen and then for the head and winter (Plate XXIV). At first he used an iron ball counterweight but this gave way to springs when these were made in sufficient quantity and strength in the United States for manufacturers such as Stansbury and Wells to take advantage of them.

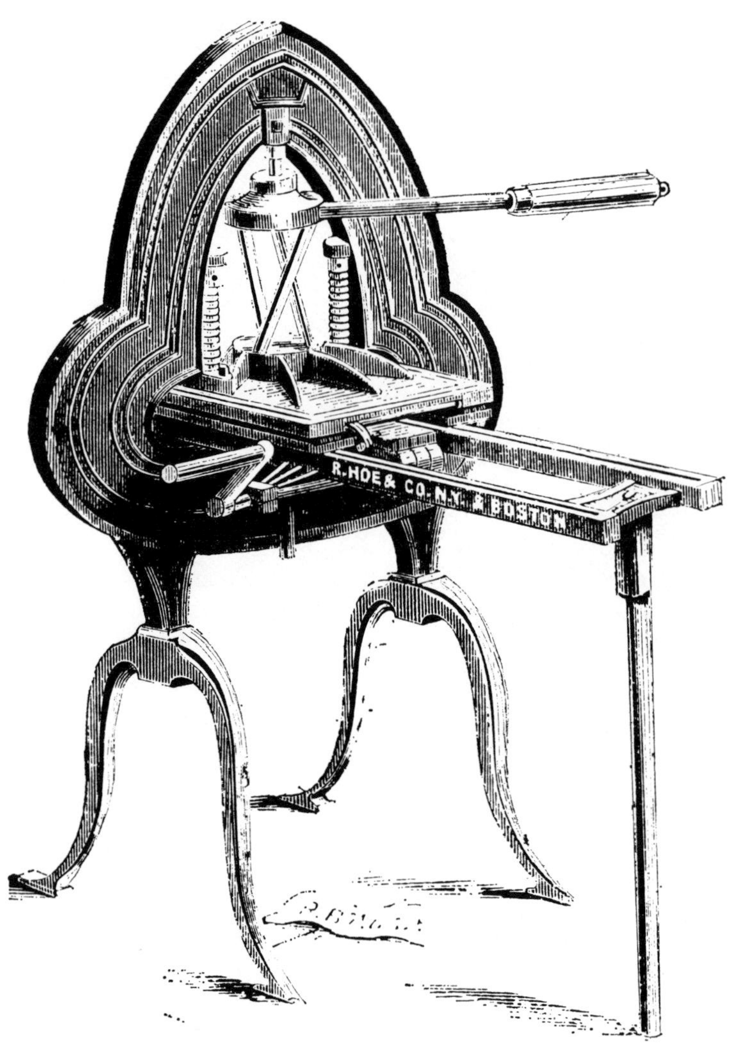

34. Hoe's small Stansbury press

Very soon after a patent was issued to Stansbury in 1821, the newly formed Cincinnati Type Foundry took over the manufacture of the press, which is listed in the typefoundry's oldest surviving specimen book of 1827. The fact that the C.T.F. referred to 'Stansbury's Patent Presses' would indicate that there was an agreement between the two parties for C.T.F. to manufacture and distribute the press on Stansbury's behalf. By 1834 the press, in medium, super royal and imperial sizes, was being described as of cast iron.

A smaller, all-iron *Stansbury*, was offered in the 1856 catalogue with a platen size of 14×18 inches. This seems to indicate that by that date the torsion toggle was considered suitable for small, cheap presses only. Larger-sized presses advertised were of the *Washington* type. This view may have been shared by Hoe's who, after the patent had run out, from 1867 to 1885, made a small *Stansbury* press with a platen of 13×17 inches. However, it had an acorn-shaped frame and the toggle consisted of three inclined rods, with their upper ends in sockets in a disc, turned by a bar. This was yet another example of Hoe's taking over an existing press and making 'improvements'.

Christian Dingler has already been mentioned as a German engineer who adapted British and American presses. He, too, produced a torsion-toggle press, with two inclined rods held at their foot in a disc. The press was called the *Hagar*, after the typefounder William Hagar, of New York (1798–1863). Why, precisely, it should have been manufactured in Germany instead of the United States is a matter of speculation. The *American Dictionary of Printing and Bookmaking* (New York, 1894) illustrates the press, and, in a brief reference, gives the impression that Hagar was a German, although there is a separate entry for the New York typefounder of that name. On the other hand, the German writer Alexander Waldow, in *Die Buchdruckerkunst* (Leipzig, 1874), definitely attributes the invention to Hagar of New York and states that the press was made by others besides Dingler.

Hagar, as was common practice among typefounders, also sold printing presses. An advertisement in the New York *Tribune*, of 3 June 1841, describes him as an agent for *Napier*, *Washington* and *Smith* presses. In an advertisement in the *Printers' Circular*, of March 1873, his successors are found offering 'power, hand and job presses of all popular manufacturers'.

Why did Hagar not arrange for the press bearing his name to be made in the United States and sell it there? The reason may have something to do with his reputation. In his *History of Typefounding in the United States*, David Bruce says of William Hagar: '. . . a man who was noted for nothing so much as a want of sincerity. In swindling the inventor David Bruce out of his patented rights, he succeeded not only in destroying his prospects but overreached himself—a peculiarity which pervaded all his transactions.' Bruce may have been prejudiced, but if Hagar obtained the rights to the Bruce typecasting machine in an improper way, he could also have copied the patented *Stansbury* press, and either finding nobody willing to manufacture his version or being aware that this might involve him in litigation, arranged for the press to be made in Germany, where he was in touch with the printing supply trade to which he had sold Bruce type-casters.

The torsion-toggle type press was thus introduced into Germany, and this helps to clear up a minor mystery propounded by J. H. Mason, a well-known figure in

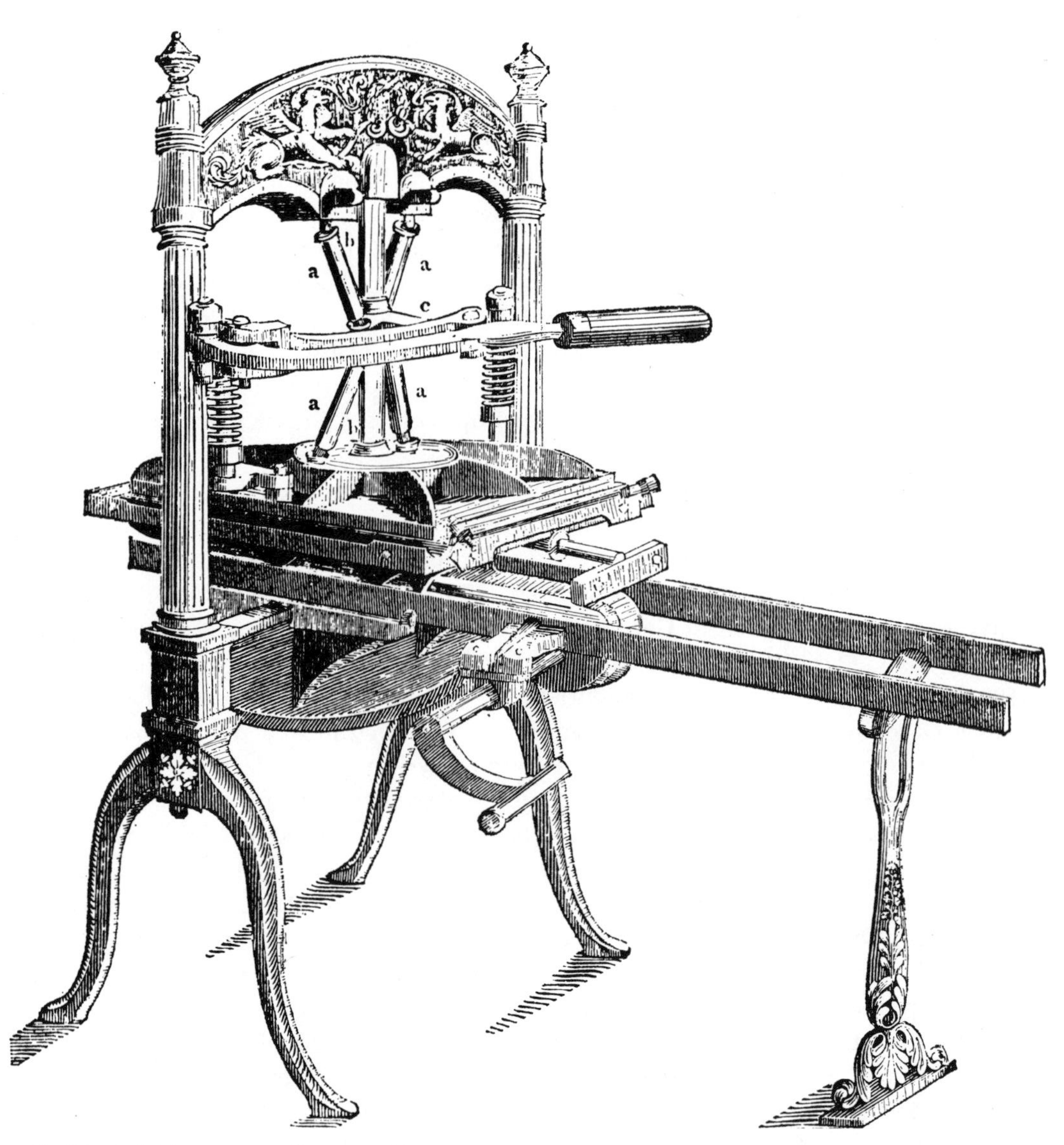

35. Dingler's Hagar press. (See page 85)

British printing during the first half of the twentieth century. In 1913 Mason went to Germany to assist Count Harry Kessler to set up his Cranach Press. In his *Notebooks* Mason explains that for equipment he visited a printing supply house in Leipzig 'where I was fortunate enough to see one of the old Medwin hand presses'. The name 'Medwin' has puzzled students of printing and it may well be that Mason's memory failed him and that he meant 'Medhurst'. His description tends to confirm this: 'The platen was operated by two oblique steel pillars, the lower ends of which were fixed to a round plate. By means of a handle the plate was moved half a revolution which brought the two bars to an upright position and

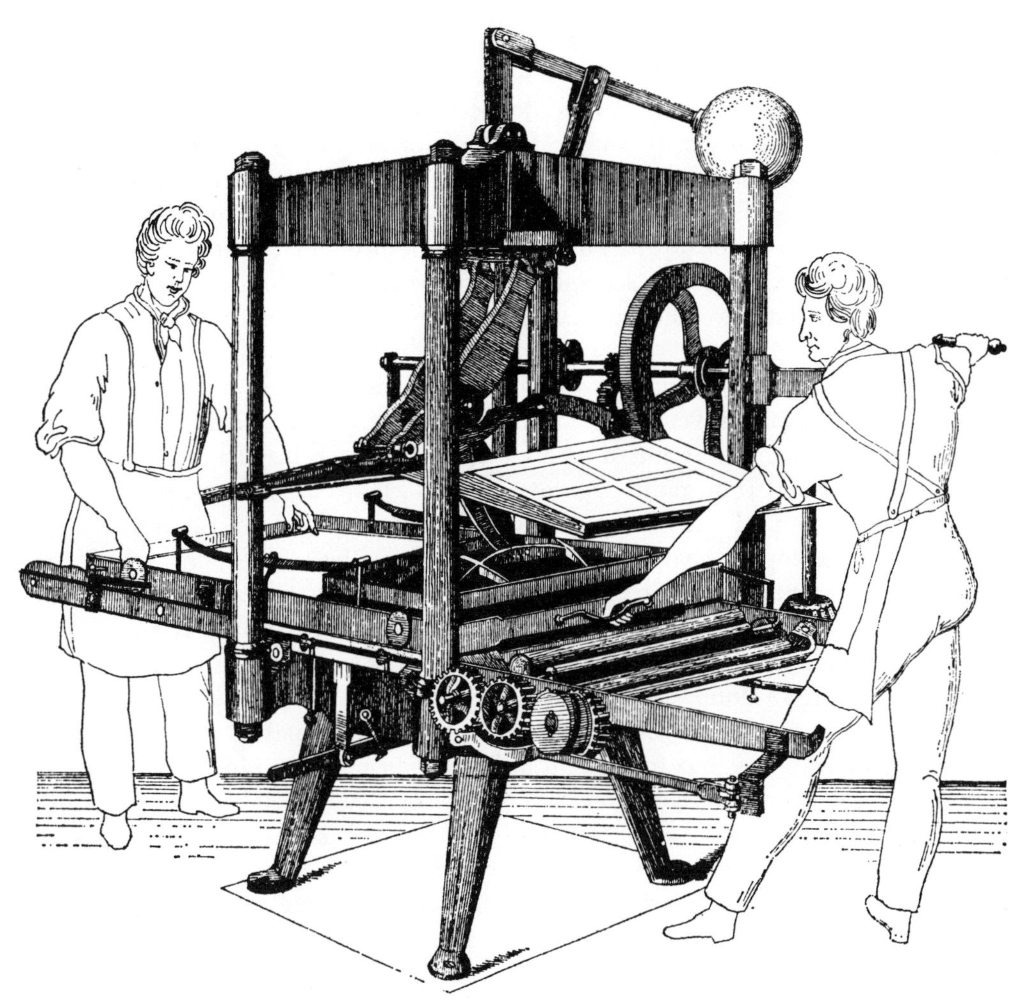

36. Deisler's version of the Selligue two-forme press

depressed the platen.' Mason states that he had never seen a 'Medwin' press in England, which is not surprising since the *Medhurst* was an adapted 'common' press (according to Hansard in 1825) and it is unlikely that a wooden press would still be in use in the printing establishments in which Mason had worked. What he probably saw in Leipzig was a *Hagar* or similar German-built torsion-toggle press, such as that made by Krause, of Leipzig.

Basically, the iron hand press was worked in the same way as its wooden predecessor, but inventors were reluctant to abandon two possible improvements—printing from more than one forme and automatic inking. Both John Brown (see Chapter 2) and George Clymer (Chapter 4) had taken out patents for presses to print from more than one forme, but there is no evidence that they were ever built. In 1827, however, a mechanic named Selligue did develop a two-forme press in Paris, which was eventually taken over by Johann Deisler, of Coblenz. Why its manufacture was transferred is not definitely known, but during the 1830 *coup d'état* in Paris, hand pressmen took the opportunity of breaking up the detested printing machines, and this may have been a contributory factor.

Whatever the case, it is Deisler's description and suggestion for use which is of interest. He described the *Selligue* press as half-way between a hand press and a printing machine; and said that this 'double press' was meant for those printers whose output did not justify the installation of a Koenig and Bauer cylinder machine. Two men operated the *Selligue* press. The first worked with one tympan and forme, carrying out the printing operations, while his companion prepared to work from a forme lower down. Turning a handle brought the second element to the printing position for the second man to work. The inking was automatic. Deisler claimed that two men in a twelve-hour working day could print 6,000 sheets.

Nothing is known of the fate of John Brown's ideas for automatic inking, but in 1825 David Napier invented a self-inking *Albion* press, which dispensed with tympan and frisket. The action of turning the rounce to bring the forme under the platen at the same time pulled round a spit a chain which extended upwards around a roller to motivate inking rollers set in a structure above the forme. Paper was fed down a sloping board to the inked forme at the point of entry under the platen. Hansard also refers to Hugh Wilson, Edinburgh engraver, who tried to make a self-inking press on the same lines as Napier; and Ruthven is said to have invented such a device for attachment to his press.

In the year 1847 John Chidley, publisher of Aldersgate Street, London, and Richard Hoe, press manufacturer, of New York, both patented an automatic inking apparatus—the first for use on a *Columbian* or other hand press, and the second for use on the *Washington*. Chidley proposed an elaborate mechanism, including an inking table at the rear of the press linked with the rounce. By turning the handle, two inked composition rollers were to pass over the forme, and by reversing the motion of the handle the forme was to return under the platen. By use of a treadle, pulleys and weights the pressman was also supposed to deliver the printed sheets on to a table below the carriage, printed side up, without having to lift the frisket by hand, the paper being delivered by the motion which withdrew the forme and frisket from under the platen. There is no record of this complicated device being adopted by Clymer, the inventor of the *Columbian*, or by any other hand-press manufacturer. If it had proved a practical proposition it would no doubt have been considered, but the very lack of information about it indicates that it was not acceptable.

Hoe's mechanism was placed by the side of the bed of the press so that when the carriage, with its forme of type, was run out from under the platen it was in a position for the inking rollers to pass over and ink the type. The apparatus was driven by a belt-driven shaft and pulleys, and the inking rollers were fed by a distributing roller from a duct of ink. Hoe's self-inker was described in the 1855 story book devoted to the large New York printing establishment of Harper's, and was mentioned as late as 1881 in the Hoe catalogue, indicating that there must have been some demand for this addition to the hand press long after the jobbing platen had developed. In general, however, it is probable that these inking devices were either difficult or too elaborate to operate and that pressmen preferred to ink the type manually. Most descriptions of hand-press printing, photographs and personal reminiscences indicate that this was so.

During the nineteenth century a number of types of press were being used commercially alongside each other, including, to some extent, wooden presses. Use was determined by economic factors and the kind of work a press was expected to produce. In the long run, the hand-inked iron press survived as a poster or proofing press, or for the limited requirements of private individuals. For small commercial work it was superseded by the more-efficient jobbing platen, and for bookwork by the 'bed and platen' press, which, in its turn, was finally superseded by the cylinder machine.

Although the cylinder machine was invented by Koenig after the *Stanhope* press had appeared it preceded in time the other hand presses and the so-called bed and platen press; and thus, from a strictly chronological viewpoint it should have been dealt with following the Stanhope press. However, it was thought more appropriate to describe each of the key hand presses directly after the first of their number, and before going on to Koenig's important invention it is thought desirable to devote a chapter to perhaps the most ubiquitous of all hand presses—the *Albion*, and its imitators.

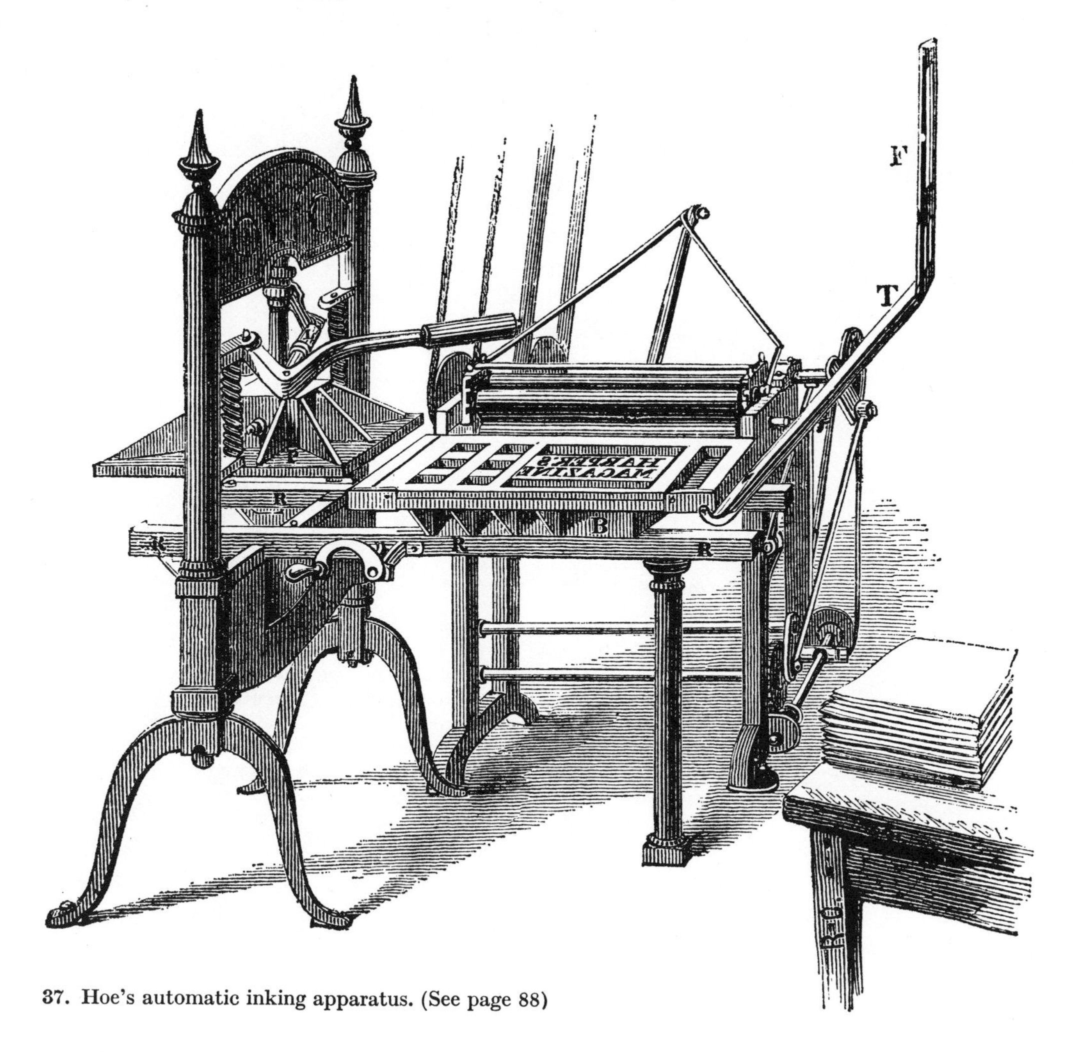

37. Hoe's automatic inking apparatus. (See page 88)

A SIDE VIEW

OF THE

Self-Inking Albion Printing Press,

INVENTED BY D. NAPIER.

Manufactured by BAISLER & NAPIER, *Engineers,*

No. 15, LLOYD'S COURT, CROWN STREET, SOHO, LONDON.

B. & N. in respectfully recommending to the notice of the Trade their *Self-Inking Press*, would briefly mention that the Tympan and Frisket are dispensed with, and consequently the labour attending them. That at the rate of five sheets per minute by two men at the *Common Press*, one man with this Press will print six; and it occupies much less space than any other Press whatever. The *Albion Press* may be had without the *Self-Inking Apparatus*, which may be applied at any future time;—and also to the generality of Iron Presses now in use. The want of room will not permit a description by the letters to be seen on the Engraving. As it respects its power or ease in labour, it is allowed by some of our best mechanical gentlemen, and confirmed by some years' experience, to be the best hitherto offered to the Trade. B. & N. being also the Manufacturers of *Rutt's Patent Cylinder Printing Machine*, and *Treadwell's Patent Foot Lever Printing Press*, beg leave to state that agreeable to the rate of Printing above-mentioned, one Pressman, a Labourer, and a Boy, with Rutt's Machine, will print from fifteen to eighteen sheets per minute; with Treadwell's Press, two men will print seven sheets per minute, with much less labour than any other common Press; and from the extreme simplicity of its construction, and comparative cheapness, is admirably calculated for exportation. Respectable references will be given where the several Machines and Presses may be seen at work.

P. S.—B. & N. beg also to inform Gentlemen in the Printing Business, that they have invented a new and superior Inking Table and Duct for the Common Press, which from its simplicity they can sell at less than half the usual price of those which are called patent.

38. The Albion press was invented by R. W. Cope but this did not prevent imitations being marketed. The above announcement is of unusual interest since it also contains information about the Rutt and Treadwell presses, as well as David Napier's method of applying automatic inking to the Albion press (see page 88). Napier was in advance of Hoe, who retained the tympan and frisket, but automatic inking devices did not find favour and hand presses are usually manually inked to this day

6

THE ALBION PRESS

WHETHER OR NOT Bigmore and Wyman's statement that Richard Whittaker Cope assisted George Clymer in the manufacture of the *Columbian* press is accurate, their further assertion that he invented the *Albion* press at his new premises in New North Street, Finsbury, is acceptable, although we have only their word for the date of the invention—1820.

An advertisement by Cope in Pigot's *London and Provincial New Commercial Directory*, published in 1823, confirms the fact that Cope moved from Bowling Alley, and that he was the inventor of the *Albion* press, a cut of which appears at the head of the page. While Hansard in 1825 provides only a perfunctory reference (not mentioning the name *Albion*), John Johnson, a year before, had given a full description of the working parts of the press and included an engraving showing an *Albion* press and an hydraulic book press, boldly labelled 'Cope', which had been mentioned in the Pigot's *Directory* advertisement.

Little is known about Cope, but a catalogue of Hopkinson and Cope, dated 1862, issued from 'Albion' Works, New North Street, claims that the establishment had been in existence for 'nearly fifty years', which can mean that Cope set up as a printers' engineer in about 1813, and would have been conversant at least with the *Stanhope* press. Whether he had been pursuing the notion of his own iron printing press before Clymer's arrival in Britain is not known, but it seems to have stimulated him into activity, and he was ready with his press in the very early 1820s.

The name *Albion* was by no means novel in the printing trade, but it was an obvious British riposte to the American *Columbian*. However, Clymer's ideas of elaborate decoration were not followed and the adoption of the royal arms as a counter-attraction to the American eagle was a much later afterthought.

It has proved impossible so far to confirm Bigmore and Wyman's date of 1820. It is also not clear what steps Cope took at the outset to manufacture his press. It does not look as if he had firm views on its appearance or about the form of mechanism for returning the platen after printing. A French import license, dated 22 November 1822, 'pour une press d'imprimerie . . . dite presse Albion', carries the words 'Presse d'imprimerie par Mr. Dunne'. A diagram of the press shows a coiled spring in a cap for returning the platen, which is supported by three bolts only. There are drawings of alternative methods of returning the platen, including a leaf spring, which suggests that other designs were considered.

A surviving press bears the name 'Cope' stamped on the bar as well as a date, the last figure of which is barely legible. It could be 1829, 1823, or 1820. The brass cap to house the spring is unfortunately missing. The platen is supported by three bolts, and as the press is a table model it has short legs terminating in claw feet. Another surviving press, with no date, but with 'Cope/London' engraved on a 'pepper-pot' brass cap, as well as being cast in the head, also has three bolts to hold the platen instead of the later four. The legs, while in a curling pattern, are not very massive and are undecorated. Two post-Cope *Albions* in the Lindner collection in California, dated 1829 and 1832 respectively, both have legs of this kind.

Most interesting, however, is the fact that cast in relief in the head of this second survivor is a small eagle with outstretched wings between two leaf patterns. This may be a faint indication of a connection with Clymer. The side guides for the piston are also cut in a manner reminiscent of the *Columbian*. Johnson's 1824 engraving by 'Mosses' shows a brass cap and rather plain legs. The Cope 1823 advertisement shows a 'bound leaf' iron housing for the spring, and legs decorated with a leaf pattern, terminating in claws on square blocks, and joined with curved pieces of metal bearing the words 'Cope/London'. The 'bound leaf' seems to have been favoured for the larger-sized presses and the brass cap for the smaller. The curling legs eventually gave way to the more substantial ones terminating in claw feet.

As far as his early presses were concerned Cope was not influenced by Clymer or by other American inventions. His toggle was of a different kind and he used a spring instead of a counterweight. It is curious, therefore, that towards the end of his life he imitated Clymer and began to use a counterweight. His successors in business continued with the counterweight briefly, as no doubt a number of presses so equipped were in stock at Cope's death, but they soon reverted to the spring as a means of returning the platen to its raised position. Perhaps Cope was feeling the competition of the *Columbian* and thought that popularity could be gained by a decorated counterweight bar and easily recognized weight, which could be seen to be working, rather than a hidden spring. If so, he was mistaken, as any lack of popularity of the *Albion* derived from a weakness in the brass links which joined the piston to the spring bolt rather than the actual means of returning the platen.

William Savage, author of the *Dictionary of the Art of Printing* (1841), was among those who complained about the breaking of the links. Only some 200 presses were sold by Cope, according to the 1862 catalogue, and it can be fairly stated that the *Albion* press was not firmly launched on its successful career until after Cope's death and until a foreman in the works, John Hopkinson, removed the links and altered the toggle mechanism.

The earliest American toggle presses utilized long levers joined together, but Cope, quite possibly independently, devised a more compact mechanism enclosed inside the piston, Johnson claiming that this was the first time this had ever been done. The piston, which slid up and down, had a rounded base which pressed into the centre of the platen. Within the piston there worked a fulcrum or mouthpiece. This was a piece of iron with an elongated hole terminating in a half-circle, in which there rested a hanging lug attached to the bar. The top of the fulcrum was cut in another half-circle to work against the trunnion protruding from the centre of the head of the press. When the bar was pulled the lug moved into an upright position, pressing the fulcrum into a similar position, and hence pushing down the piston.

In about 1827 Cope decided to replace the spring with a counterweight on a bar decorated with a trellis pattern. On a surviving press, No. 185, dated 1827, the counterweight is in the shape of an urn. Apart from the urn, this press resembles an undated engraving inserted in a copy of William Savage's *Practical Hints on Decorative Printing* (1822), in the St. Bride Printing Library, in that it has a screw at the base of the piston for squaring up the platen. Another survivor, No. 234, dated 1828, is again similar, but the urn has been replaced by a fine, elaborate casting of the royal arms of the period. The fulcrum is also fitted with a wedge, the purpose of which is to adjust the impression. Instead of the platen-adjuster the whole base of the piston and attachment to the platen has been simplified. Similar features are characteristic of the *Imperial* press, made by a firm called Sherwin and Cope, but the improvement to the 'mode of adjusting the pull' was claimed by

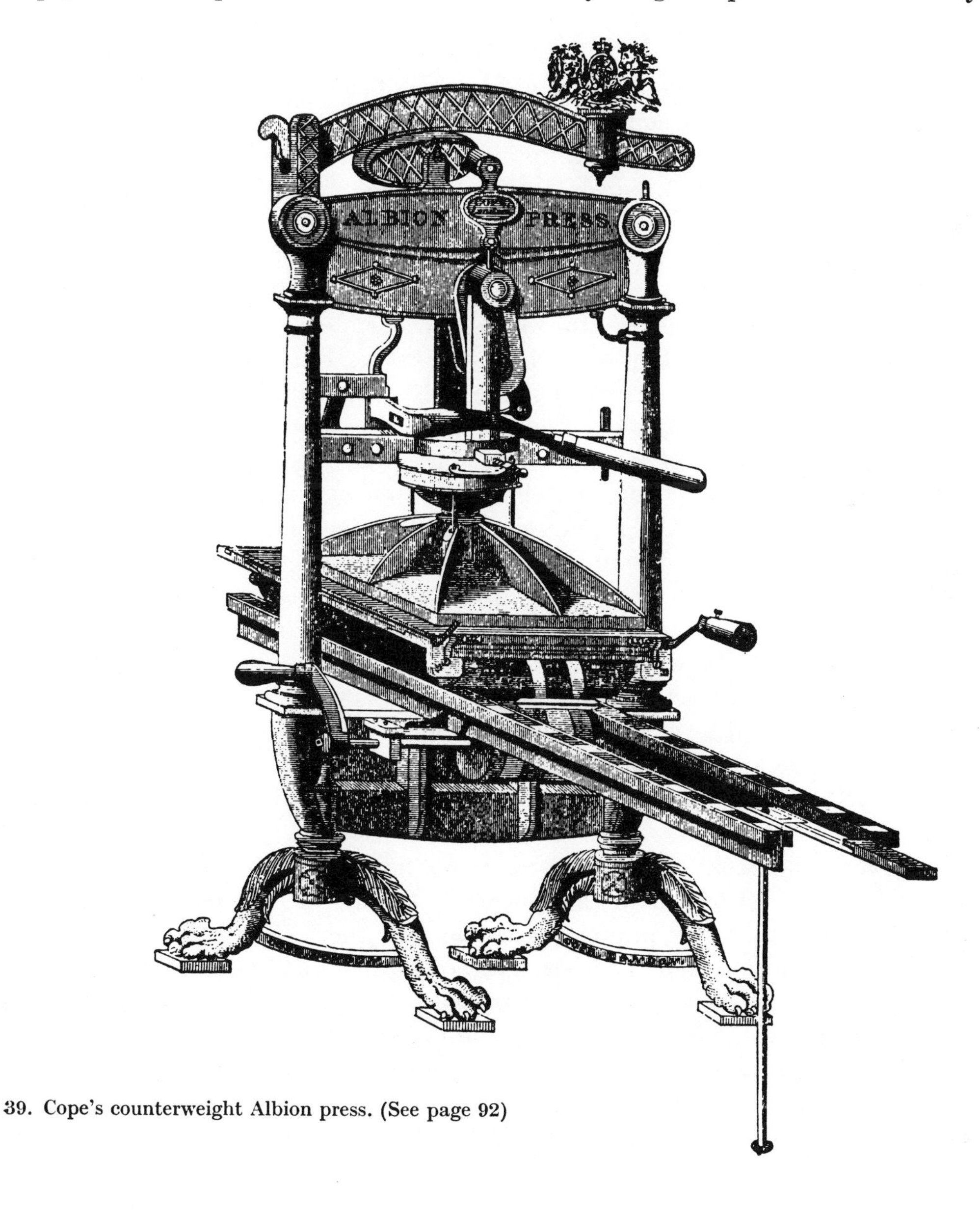

39. Cope's counterweight Albion press. (See page 92)

John Hopkinson, who in a letter to the *Mechanics' Magazine* stated: 'I introduced it in the late Mr. R. W. Cope's *Albion*, before Messrs. Cope and Sherwin's press was thought of. . . .'

Cope must have died before the date of publication, and the wording of the letter indicates that he could not have been the Cope of Cope and Sherwin. In fact, the date of his death can be fairly closely established. Press No. 234, of 1828, bears his name, but another similar press of the same year carries the description: 'Albion Press manufactured by the Exors. of the late R. W. Cope, London'. Cope therefore died in 1828, and before the middle of the year, as in an old account book of Messrs. Richard Clay there is an entry, dated 30 August 1828, referring to 'Cope's Execrs.'

Hopkinson, who took over the management of Richard Cope's firm, is credited with the improvements to the *Albion* press in literature put out by the firm, and he may be credited with priority in the case of the impression adjuster. However, two other changes, which proved to be important to the *Albion*, the abolition of the links and the use of a chill (which will be described later), owed much to Messrs. Sherwin and Cope, of Cumberland Place, Curtain Road, Shoreditch, London, makers of the *Imperial* press. A description of this press in the *Mechanics' Magazine*, dated 26 September 1829, to which Hopkinson took exception, refers to the *Imperial* as a new press, and a prospectus, dated 1832, in the John Johnson Collection in the Bodleian Library, Oxford, shows a press dated 1828.

It is not known whether J. Cope, of Sherwin and Cope, was related to R. W. Cope, but he probably was, as it seems too much of a coincidence that a man of the same name should have brought out a rival press in the year of R. W. Cope's death. He was not R. W. Cope's son because Bigmore and Wyman, in the oft-quoted passage, wrote: 'On the death of Mr. Cope the business was carried on by his executor, Mr. J. J. Barrett, a brother of the printer of Mark Lane, the style of the firm being Cope & Barrett. Their foreman, Mr. John Hopkinson, an admirable man of business, was soon introduced into the partnership, and carried on the trade with honourable fidelity and great success during the long minority of Mr. Cope's son. . . .' J. Cope may not have seen eye to eye with the executors, who were acting on behalf of the infant son, and may have taken his knowledge and experience elsewhere.

Whether that was so or not, the *Imperial* press used mechanisms which were later incorporated in the *Albion*. The *Imperial* had no links and no spring in a cap—the platen was raised by a leaf spring. Stanhopean levers were linked to the toggle which was in the head of the press. In consequence, it was a more powerful press than the *Albion*, and was often converted into an arming press for binders and is still occasionally so used.

Whoever should take the credit for the abolition of the links and the new-style fulcrum, the *Albion* benefited. Instead of the simple fulcrum and lug there was, from 1830 onwards, a split piston, to accommodate the adjustment wedge, and the lug was replaced by a chill. This was a steel bar with rounded base, working in a trunnion at the top of the piston, and its own head inside the rounded part of the fulcrum, a movable attachment of the piston. When the bar, which is connected to the fulcrum, is pulled, the fulcrum moves to a vertical position, forcing the chill to do likewise and thus thrusting down the platen. This form is not followed in all Albion-type presses. In some, the later Hopkinson and Cope versions, for example, the hollow fulcrum is dispensed with, and in small presses there is no adjusting wedge. In others the wedge is inserted at the base of the piston.

Cope had, in fact, two executors, Jonathan and Jeremiah Barrett, and after 1830 they had their names cast on the head of the staple and as a sort of compromise with the practical man of the works, John Hopkinson, 'Hopkinson's Improved Albion Press' was engraved on a brass oval or cap. They also used the word 'patent' on the piece of metal which connected the bar to head and staple and on top of this a casting of the royal arms (Plate XXV). Neither the word 'patent' or the arms were used with authority but were presumably to persuade buyers that the press was of an original, protected design.

The executors continued to have their name on the press well into the 1840s, although Hopkinson was acknowledged as the manufacturer, completing his 1,000th improved press in April 1839. By 1847, at least, the name on the press was Hopkinson and Cope. Hopkinson died on 16 October 1864, aged 67 years, and the business was taken over by the former infant, James Cope.

The word 'patent' and the coat of arms on the *Albion* did not deter other firms from making (or having made) and distributing their own versions of the press. As in the case of the *Columbian*, once the *Albion* became an established favourite there were many wanting to share in the profits. The firm of Hopkinson and Cope did not go under as the result of the competition, as it was able to move into the printing-machine field and so, besides manufacturing 'Hopkinson's Improved Albion Press', it was also responsible for the *Scandinavian* printing machine, which it had taken over from another firm; a double platen of its own invention; the

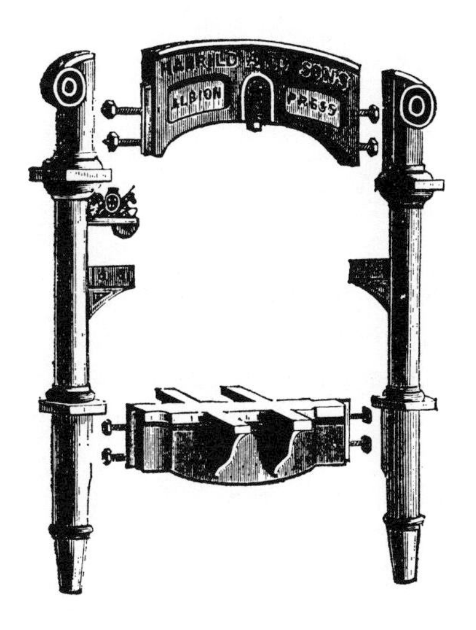

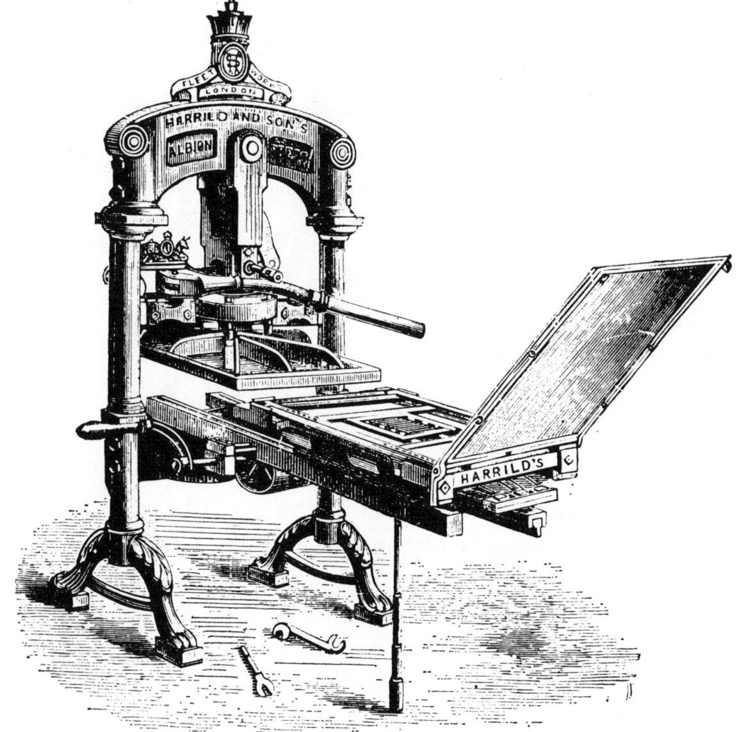

40. Harrild's portable overland Albion press. (See page 96)

Anglo-French machine; a single-cylinder machine; a beam engine; and a wide range of printers' and bookbinders' equipment, including 'Steam *Albion* arming presses'. By 1871 it was supplying *Imperial* arming presses. Whether it had taken over Sherwin and Cope or had simply appropriated the name *Imperial* is not known, but about this time it joined forces with Payne & Co., and was eventually absorbed into Dawson, Payne and Lockett.

As in the case of the *Columbian, Albion* presses were supplied by all the major printers' supply houses—Harrild and Ullmer, for example, and by the typefounders, such as Miller and Richard, Caslon and Stephenson Blake. Other names appear on presses—Notting, Morton, Hughes and Kimber, Sherwood and D. and J. Greig, of Edinburgh. How many were actual manufacturers it is impossible to say. Thomas Matthews, of Cow Cross Street, Smithfield, who started up in 1848, confidently claimed to be a manufacturer, and called his works by the name of 'Albion'.

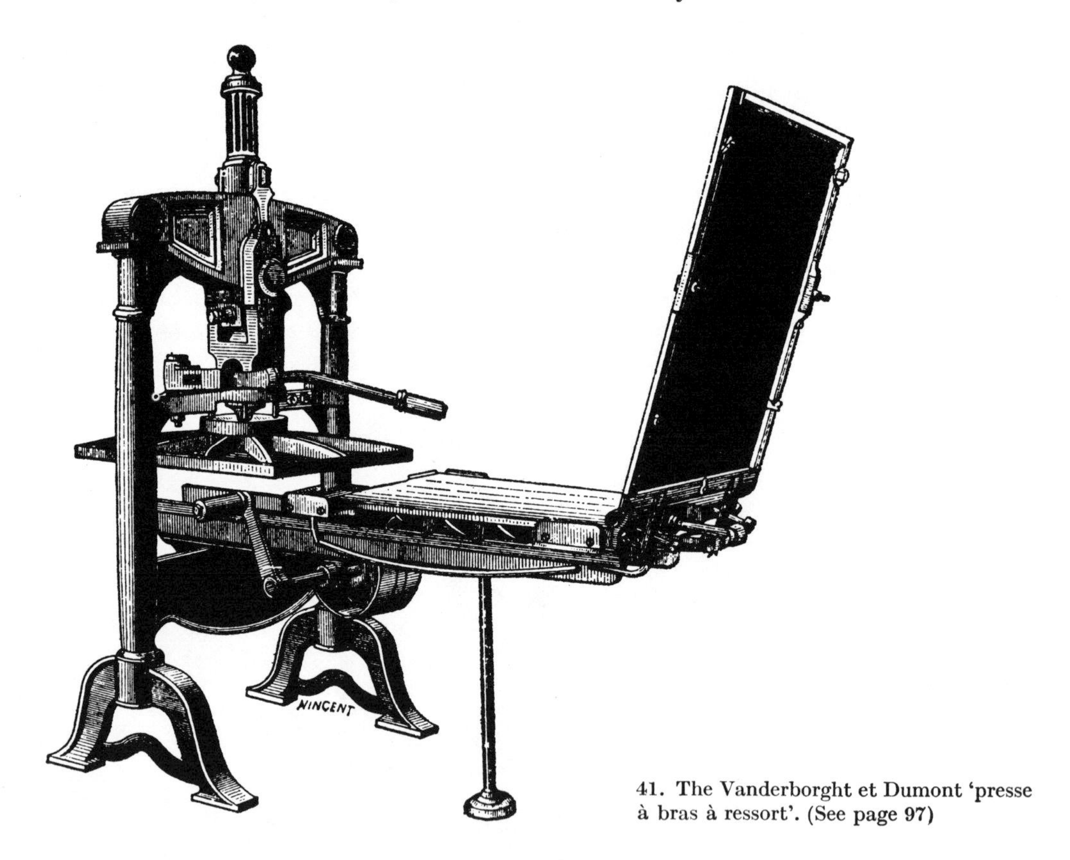

41. The Vanderborght et Dumont 'presse à bras à ressort'. (See page 97)

Samuel Rust's idea of a press which could be taken apart for transporting over distances in the United States (see Chapter 5) was taken up by Harrild's on behalf of the British Empire. They advertised a portable overland *Albion* press 'extra strong, with mortised staple made in six pieces for ease of transport to up-country districts abroad'.

A Lockett's undated list shows that *Albions* were made in sizes from 'Amateur', $7 \times 5\frac{1}{2}$ inches, to Double Royal, 40×23 inches. The royal octavo, $10\frac{1}{4} \times 7\frac{3}{4}$ inches, was usually designated a 'card' press. However, there is an existing miniature

Albion-style press (no manufacturer's name) with a platen as small as $6 \times 4\frac{1}{8}$ inches; it is only $18\frac{1}{2}$ inches high. Hopkinson and Cope, from their 1862 catalogue, made nothing smaller than the foolscap folio (called 'half sheet foolscap'), $15 \times 9\frac{3}{4}$ inches. This was priced at £12 10*s*. The price of the Double Royal was £75. They also made two *Albion* 'galley' presses—one 29×6 inches (£12 12*s*.) and the other $36 \times 7\frac{1}{2}$ inches (£18).

Overseas manufacturers issued their own versions of the *Albion*, although English presses were exported. The firm of Gottl. Haase Söhne, of Prague, advertised small English *Albions* in an 1845 issue of the *Journal für Buchdruckkunst*, but five years later Laurent et Deberny, of Paris, advertised their own *Albion*-style press with a lamp-shaped counterweight. Cope's counterweight *Albion* was copied fairly closely by the Belgian firm of Lejeune, even down to the trellis pattern on the counterweight bar. But a Belgian lion was substituted for the coat-of-arms counterweight. The Vanderborght et Dumont 'presse à bras à ressort' was inspired by the spring-top *Albion* and in Australia, F. T. Wimble & Co. of Melbourne and Sydney, manufactured an *Albion* press (patent 1385) and a Crown size survives, dated 1879.

Perhaps the most extraordinary example of an *Albion* is a small, roughly foolscap folio, press, which survives in Singapore, which, while following the original design is unlikely to be of English manufacture (Plate XXVII). There is a Japanese or Oriental air about it. The Japanese would not have manufactured presses before the late 1870s, but it may be that an enterprising manufacturer thought he could capture the Eastern market with this variation of the *Albion*.

Dawson, Payne and Elliott, of Otley, Yorkshire, continued to make *Albions* as late as 1940. Four Super Royal presses were ordered in that year by the India Office Stores department, Bangalore.

Other presses based on the *Albion* were put on the market. W. H. Lockett & Co., following the patriotic tradition, named their press the *Alexandra*, in honour of Princess Alexandra of Denmark, who married the Prince of Wales in 1863. The cap was consequently in the form of ostrich feathers from the Prince of Wales badge. An *Alexandra* was made by or for the Australian firm, F. T. Wimble & Co. (patent no. 1355), and a double-demy size, dated 1878, survives.

W. Notting built an *Alexandra* press, and the fact that it differed very little from his *Albion* indicated that manufacturers were often concerned more with publicity and fashion than with technical change when they issued a new version of an old press. The popularity of the *Albion* not only led to widespread imitation but also to a general mix-up of design elements and a fine confusion of nomenclature. For example, J. Smith, of Denmark Street, Soho, brought out an 'Improved Hercules Press', based on the *Albion*, but with square-section fluted columns and curling legs. Walker Brothers, of Bouverie Street, made an 'Improved Albion Press' with columns and the Prince of Wales's feathers obviously copied from the *Alexandra* and with leaf patterns in high relief at the shoulders of the staple inspired by Notting's version of the press.

The *Lion* press, clearly inspired by the *Albion*, was manufactured or built for Frederick Ullmer, of 15 Old Bailey, in the 1860s—the earliest advertisement so far traced in the *Printers' Register* being dated 6 December 1866, in which it was described as 'Frederick Ullmer's new Lion Press'. It was made specially for

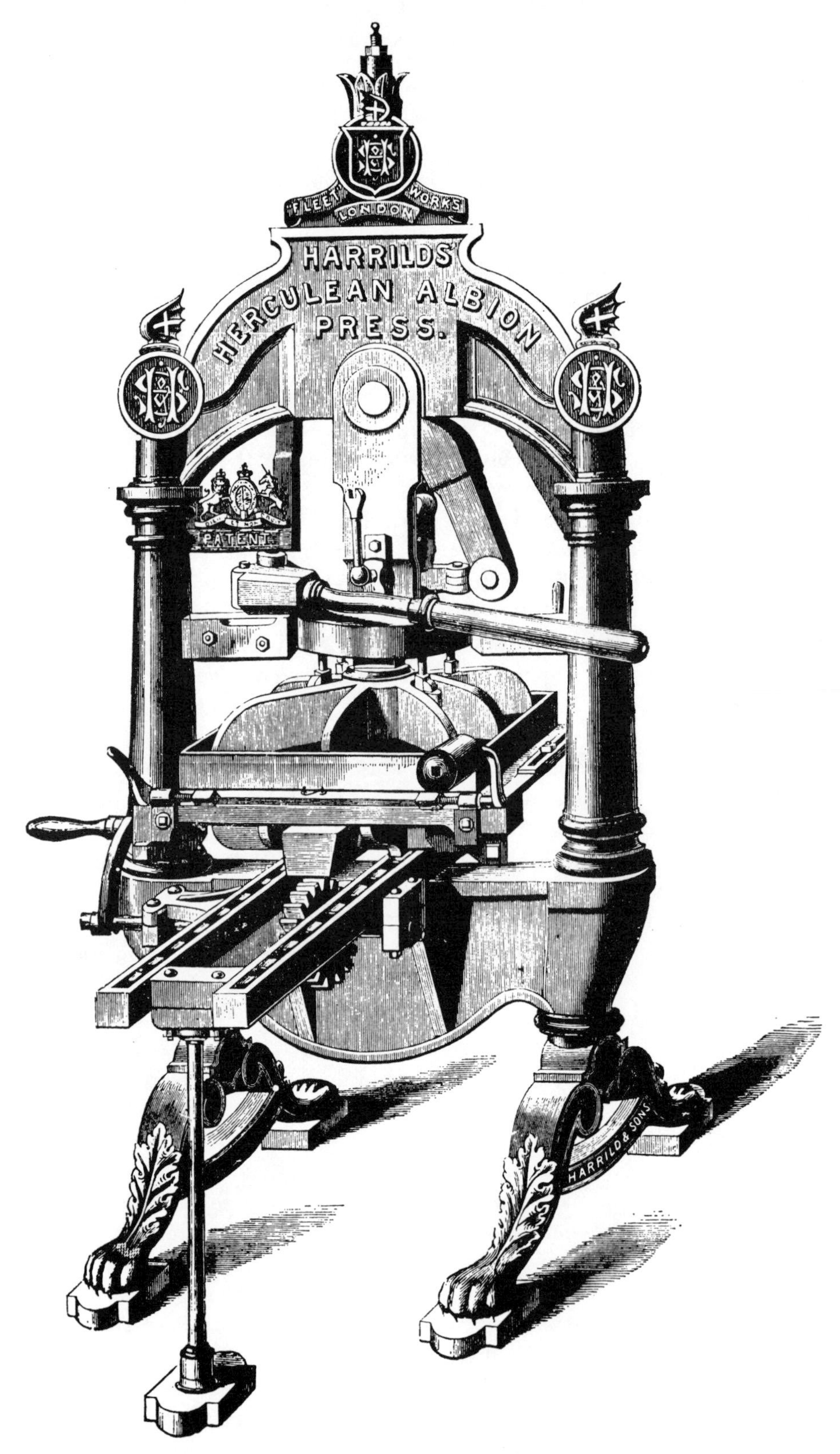

42. The Harrild Herculean Albion press with rack and pinion in place of the traditional spit-and-strap mechanism. (See pages 97 and 99)

embossing, bank-note printing or any class of work requiring immense power of impression. Only one size was available—demy, with a platen of 24×18 inches—and the price was £60. The advertisement carried a slogan in which it was claimed that thirteen *Lion* presses were in use in one office.

Ullmer was a general supplier of printers' equipment, both new and second-hand, and may have found it difficult to sell his own press at £60 when he had similar-size *Albions* and *Columbians* available at a lower price. An advertisement of 6 February 1867 quoted a new *Lion* press at £60 and a nearly new demy *Albion* at £29. He rarely advertised the *Lion* press on its own in trade journals after its initial launching, although the advertisement appeared in his 1891, 1900 and 1902 catalogues, with the word 'thirteen' altered to 'fifteen' in the slogan. It does not look as if the *Lion* press was very successful, and one of the more compelling reasons for this may have been the competition from an exactly similar press called the Harrild *Herculean Albion*, specially built for proofing and of exceptional strength. Like the *Lion* it had a strong rack and pinion for winding the bed under the platen instead of the traditional spit-and-strap mechanism (Plate XXIX).

An examination of the *Lion*, the Harrild *Herculean* and the Hopkinson and Cope *Albion* presses of like size reveals an identically cast frame, including the impressive but misleading cast of the royal arms and the word 'Patent', and this strengthens the impression that these presses all came from the same foundry, even if the head-pieces differed.

Two surviving *Lion* presses, one in the London College of Printing (Plate XXVIII) and the other at the Vintage Press, Molalla, Oregon, U.S.A., both have the Ullmer name removed and that of 'J. Esson' bolted in its place. This seems to underline the difficulties Ullmer had in disposing of his *Lion* presses, and he must have sub-contracted some of them to John Esson, a well-known nineteenth-century printers' engineer and valuer of Fetter Lane. Esson had a fairly extensive business since he had a warehouse and showroom in other parts of London, and a showroom in Liverpool. He sold all kinds of presses and his name is cast on a number of them.

Ullmer followed the patriotic tradition too, in a way, when he named his press the *Lion*, and surmounted it with a crown, but Harrild appealed to another notion—the idea of strength—when he called his press the *Herculean*, although acknowledging that it was basically an *Albion*. This line of thought was pursued in the *Atlas* press, issued by Wood and Sharwood in about 1840, and by at least one other firm—Coniston and Smith, of 101 Bunhill Row. A figure of Atlas, holding up the world, is placed in front of the cap (Plate XXVI). However, the printing mechanism is not based on the *Albion* toggle-joint system but reverts to the earlier idea of wedges, or inclined planes and rollers, in the manner of Barclay's press (see Chapter 5). Barclay relied on a wedge working between single rollers, but the *Atlas* printing unit consists of a wedge pulled by a bar between a pair of rollers above and below.

Albions and their derivatives tend to be associated with the private press movement since they were used by Dr. Daniel at Oxford, William Morris at the Kelmscott Press, St. John Hornby at the Ashendene Press, Cobden-Sanderson at the Doves Press and scores of followers all over the world. A photograph of the interior of the Doves Press (1900–16), indeed, shows a *Harrild* press, with the outlines of the *Herculean*, but without its decorative elements.

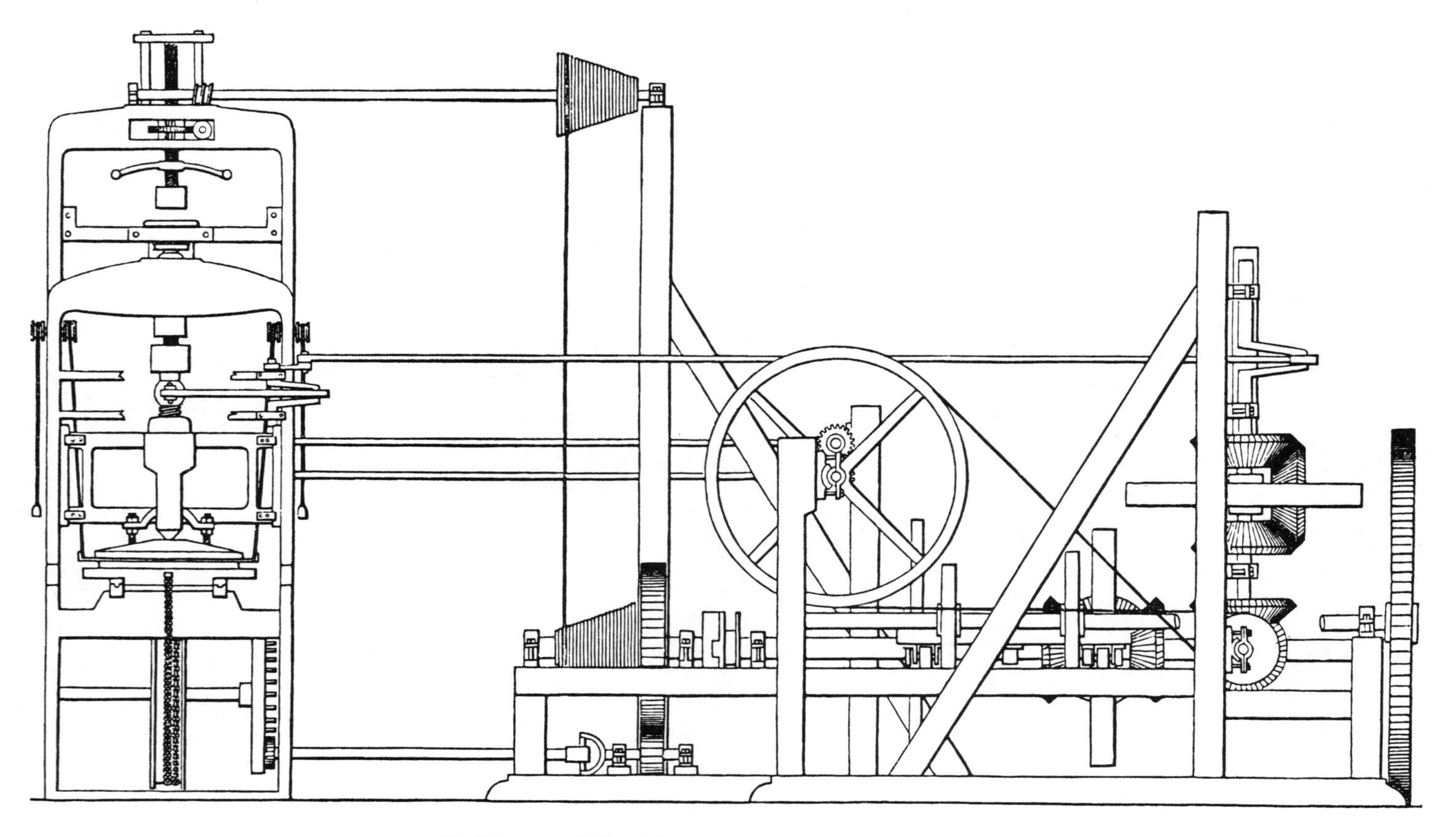

43. Diagram of Koenig's 1810 steam-driven platen machine

7

KOENIG AND THE CYLINDER MACHINE

THE IDEA THAT a rolling cylinder might be used in printing to overcome excessive manual strain was put forward as early as 1616 by Faustus Verantius in *Machinae Novae*, published in Venice, with descriptions in Latin, Italian, Spanish, French and German. His reference to 'A Wheel for Printing Engravings' may be translated: 'I have frequently seen that letterpress printers as well as those who print the multitude of engraved copper on to paper do this with great strain and not with equal result all the time, but now better now worse, on account of the variability of the equipment and unevenness of the strength. Therefore I have invented for this purpose a wheel which is most useful for such effort, which can also be driven by a boy, and that it also prints the pictures all the time in the same way.'

While this primitive invention was primarily intended for printing from engraved plates with what looks like a heavy mill-stone, theoretically it could also have been applied to printing from type, but, as will be noted, a cylinder of less weight and less crushing power was ultimately used in letterpress, or relief printing.

Before attempting to deal with this, one of the most important technical developments in the history of the printing press—the cylinder printing machine—it would be as well to eliminate a distracting element in the shape of 'Nicholson's patent'. Much has been written about Patent No. 1748, taken out by William Nicholson on 29 April 1790. F. J. Wilson and Douglas Grey, in their *A Practical Treatise upon Modern Printing Machinery* (London, 1888), go so far as to suggest that Nicholson should go down to posterity as the inventor of the printing machine, a statement somewhat undermined by another: 'Unfortunately, Nicholson, though not exactly a theorist pure and simple, was too little of a practical printer to turn his invention to real account.' At a meeting of the Institution of Civil Engineers on 5 April 1887, William Blades, printer and biographer of Caxton, said: 'Englishmen ought to be proud of Nicholson's genius, though not, perhaps, of his character; and no history of printing ought to be written without giving him the credit of his invention.' Unfortunately for Blades there was no invention as such.

Any objective examination of Nicholson's patent will lead to the same conclusion as that of Friedrich Koenig, the true inventor of the cylinder printing machine, that it was insufficient and superficial. Nevertheless, the question of priority of invention has been the subject of acrimonious disputes, and it is therefore desirable to touch on Nicholson's ideas even though they did not take practical shape.

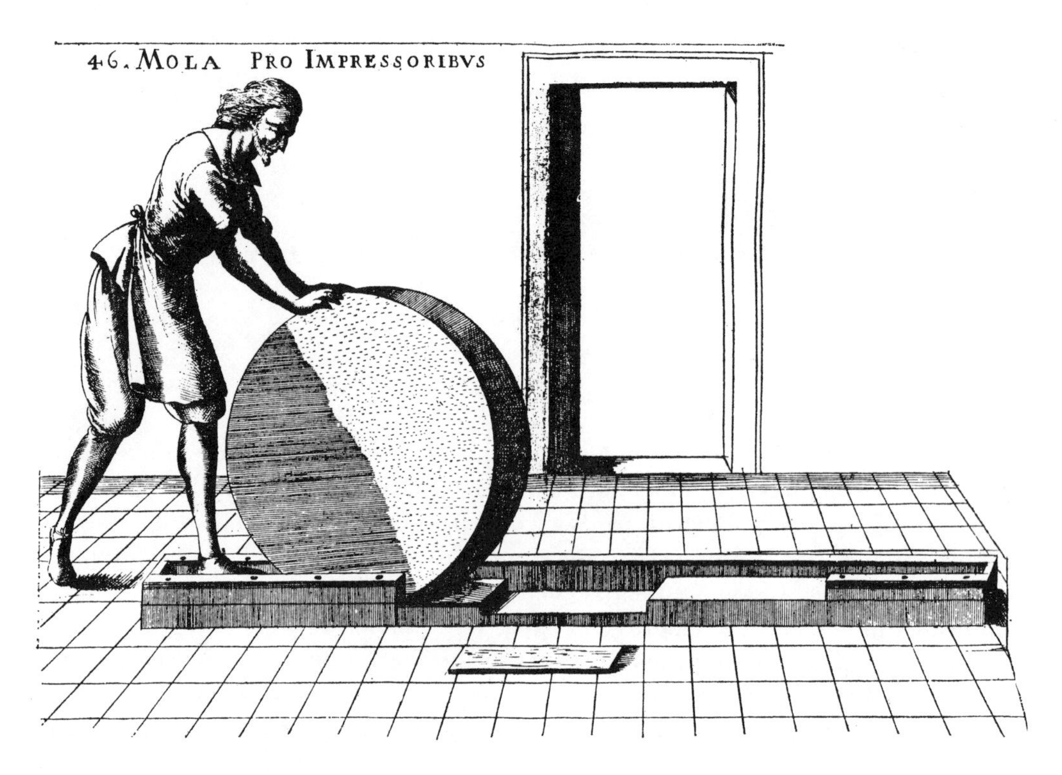

44. A wheel for printing, Venice, 1616. (See page 101)

Nicholson's patent consisted of three parts. The first was for casting types in a multi-letter mould, so that 'two, three or more letters' could be cast at one pouring of the metal, but the resulting types were to be scraped into a shape so that they could be inserted around a cylinder. The second part called for cylinders covered with leather or cloth to distribute the ink. The third demanded that all printing was to be performed by passing paper or material to be printed between two cylinders, one of which 'has the block, form, plate, assemblance of types, or original, attached to or forming part of its surface'.

The drawings of presses were of three kinds. The first was one in which the type table passed between an upper and lower cylinder; the second consisted of a printing cylinder, and geared to it a pressing cylinder below and an inking cylinder above; and the third, a press for printing fabrics of considerable length, in which the pressing cylinder was uppermost and colouring was effected by an intermediate roller from a trough roller below.

The most revealing part of Nicholson's patent in his own words consists of: 'The materials, the adjustments, the fittings, and that degree of accuracy necessary to the perfection of every machine, have likewise no part of my specification, because every workman must know that no mechanism can be completed without a due attention to these particulars.'

A few years later, in 1796, Dr. Apollos Kinsley, of Hartford, Connecticut, constructed a cylinder press on which he printed one issue of a miniature newspaper, the *New Star*. This press was referred to by Isaiah Thomas in his *History of Printing in America* (1810), but he added that it never came into use. Thomas compared it with Nicholson's patent, but stated that while Nicholson placed his formes horizontally, Kinsley placed them perpendicularly—a method 'not calculated for neat printing'. Kinsley's press was patented on November 1796, but no precise description of it exists. He continued his efforts, and is reported to have produced three other models. The description in the *American Review*, i (1801), indicates that one was based on horizontal formes, the second on perpendicular plates with type on both sides (they were to move up and down between two sets of rollers to print two sheets at once), and the third on types fixed round a cylinder, the paper to be printed being pressed between this and another cylinder. These ideas were not put into operation because of 'indisposition', but it is more likely that the lack of facilities in the United States at the time for making precision components and suitable inking rollers was the real cause of failure.

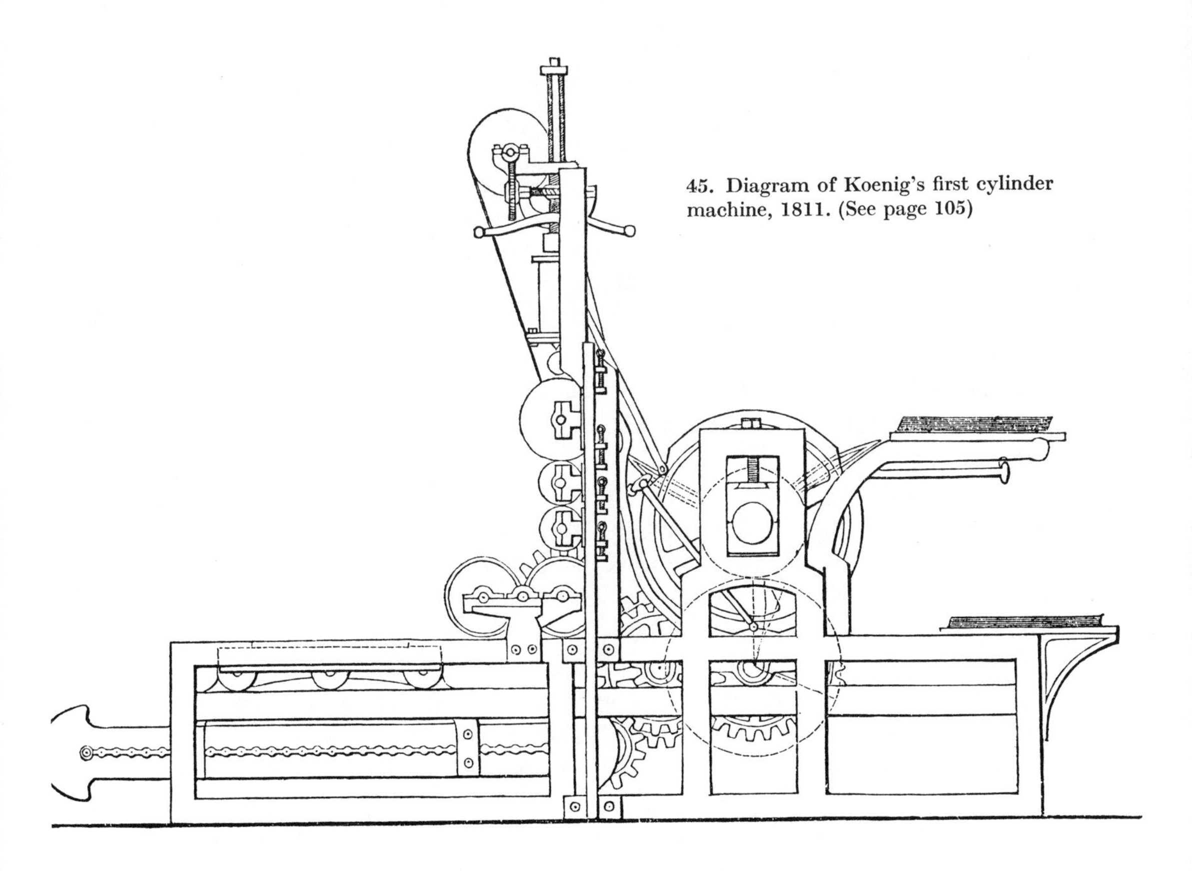

45. Diagram of Koenig's first cylinder machine, 1811. (See page 105)

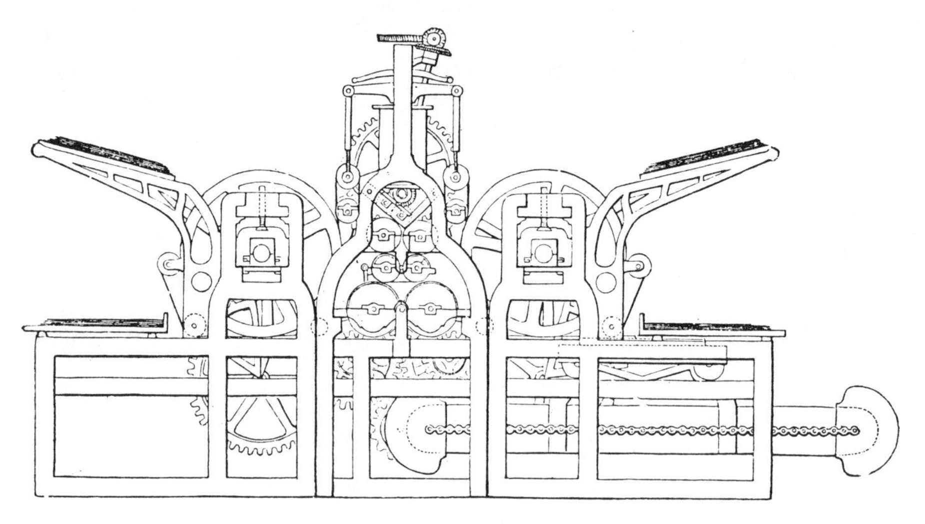

46. Diagram of Koenig's double machine for *The Times*, 1814. (See page 106)

In 1808 a M. Sutorius, of Cologne, took out a French patent for a cylinder machine, but financial ruin overtook him before he was even able to try his press out, and the patent lapsed into the public domain. Two cylinders were placed in the centre of the machine, one above the other, and the forme, tympan and frisket were drawn over four rollers and between the cylinders by the turn of a handle. Teeth placed beneath the bed at each end engaged in pinions connected to the tympan frames and automatically lowered them on to the forme.

John P. Sawin and Thomas B. Wait, of Roxbury, Massachusetts, patented a cylinder press three years later, in 1811, but its exact construction is not known owing to the destruction of the patent specification by fire in 1836. It apparently still utilized a tympan and frisket and could be worked by horse-, water- or steam-power. William Elliott, of New York, tried out a hand cylinder press in 1813 but shelved the project.

From what is known of Kinsley's first press it seems that the impression was to be made on the paper by passing the same, with the types, under a roller covered with several folds of cloth, resembling the felt in a copperplate rolling press; and Nicholson's and Sutorius's drawings look very like those of a rolling press, and this was possibly the origin of their ideas.

Whether the presses would have worked is doubtful, as the heavy revolving cylinders, which are necessary to copperplate printing to press the paper into the engraved plate, if used for letterpress printing might have damaged the type. A misconception of the nature of the cylinder has perhaps misled Nicholson's admirers. For when the practical cylinder printing machine did emerge—the brainchild of Friedrich Koenig—it carried a single cylinder, not a pair, and of quite different structure to the heavy pressing roller. It was hollow, and acting as platen and tympan, impressed the paper on the inked forme in a rapid, sharp motion and not with a long crushing effect. In a way Koenig emulated Gutenberg, who avoided this crushing effect on the domestic press by hanging the platen freely by similarly adapting the cylinder to the needs of typographic printing.

Nicholson and Kinsley may have foreseen the rotary printing machine, in which there are two cylinders, one of which carries the printing surface, but even this was a development of the copperplate rolling press used by calico printers before their time, as will be shown when the rotary press is considered. They were not pioneers of the true flatbed cylinder machine, which can confidently be attributed to Koenig, who, far from copying Nicholson's patent when he read it in 1807, went on to make a power-driven press based on the old principle of screw and platen.

Friedrich Koenig was born in Thuringia in 1774. He became a printer and bookseller, and shared with others the idea that the hand press might be mechanized. In about the year 1803, in Suhl, Saxony, he produced a design which has come to be known as the *Suhl* press—a power-driven device with inking rollers at right angles to the travel of the carriage which inked the type forme as it passed to and from the platen (Plate XXXIII). The inking rollers were to be of leather, covered in the manner of the old ink-balls. Whether an actual machine was constructed is not known, but, in any case, there was little support for Koenig's ideas on the Continent —understandably in the economic, technical and political situation of the time—so he travelled to London in 1806. Here he was introduced to Thomas Bensley, the printer of Bolt Court, Fleet Street, who was interested in the idea of printing machinery, and who brought in two fellow printers, George Woodfall and Richard Taylor, to help finance Koenig's experiments.

Bensley was not concerned with putting machines on sale if they could be successfully built, but rather to restrict them to himself and his partners. Koenig teamed up with a compatriot, Andreas Bauer, an engineer, and they applied for a patent in 1810 for a machine which was, however, still based on the platen, although steam-driven. From their plans a machine was actually built and was set to work in Bensley's office in April 1811, upon which sheet H. of the *Annual Register* of that year is said to have been printed.

The inking apparatus consisted of several cylinders vertically arranged, above which was an ink-box, through a slit in which the ink was forced by a piston to fall on the cylinders, by which it was distributed. These cylinders were perforated brass tubes, through the axles of which, also perforated, steam or water was introduced to moisten the felt or leather covering. Koenig and Bauer, unlike Nicholson, gave detailed specifications of the 'mill work' which carried the carriage backward and forward and depressed the platen. This operation was accomplished by a compound lever causing a screw to make a quarter of a revolution. The tympan was raised and thrown back, as the carriage left the platen, by a chain attached to the end, while a bar depressed it into position again as the carriage returned. The frisket, instead of being hinged to the free end of the tympan—as in the hand press—sprang up by the action of counterweights the moment the tympan was thrown back, thus releasing the sheet of paper, which was changed by hand. The press is said to have worked at the rate of 800 impressions an hour—a great advance on the hand press—but it was really a dead-end; it could advance no further technically, and the inking apparatus was considered unsatisfactory.

Koenig therefore spared no time in getting to work on what was to be the first cylinder machine, for which a patent was issued in October 1811, and a version of which was ready to work a year later. This steam-driven machine, revolutionary though it was, still incorporated vestiges of the hand press, as certain developments necessary to transform the printing press completely had not yet taken place. The forme no longer made a simple movement under a platen, rather the bed on which it was fastened received a continual motion by means of a double rack—for every sheet it moved to and fro (Plate XXXIV).

The platen was discarded in favour of a 'pressing cylinder', which was completely novel. Koenig, writing later in *The Times* of 8 December 1814, explained the difference between the earlier cylinders and his invention: 'Impressions produced by means of cylinders, which had likewise been already attempted by others, without the desired effect, were again tried by me upon a new plan, namely, to place the sheet round the cylinder, thereby making it, as it were, part of the periphery.' Koenig's machine was, therefore, not a mangle, in which a sheet is rolled and pressed, which was the essence of earlier ideas, and of some yet to come, but an ingenious device for bringing the sheet of paper rapidly to the point of impression.

In the absence of grippers, a continuous motion to Koenig's cylinder would not have allowed the feeding of sheets, so there had to be an intermittent or stop motion. The cylinder was therefore divided into three parts, which were covered with cloth and provided with points in the manner of a tympan on a hand press; and iron frames, which continued to bear the name of 'friskets', were attached to hold the sheets of paper. The surface of the cylinder between the 'tympans' was cut away to allow the forme to pass freely under it on its return. The cylinder made one-third of a revolution for each impression and then stopped. The sequence was as follows: the uppermost frisket seized a sheet of paper and moved into the next position; the sheet formerly in that position came into contact with the forme and was printed; the third segment moved to the upper position.

Composition rollers were in their infancy, and at this point Koenig utilized once again leather-covered rollers, which were not very efficient, and it was also difficult to supply them with an even flow of ink. The ink-box consisted of a vertical cylinder with a hole at the base, about half an inch in diameter, and was fitted with an air-tight piston, which was depressed by a screw which forced the ink out on to the rollers. Whatever the drawbacks of this machine, it was set to work at the rate of 800 impressions an hour. Sheets G and X of Clarkson's *Life of William Penn*, volume 1, were the first ever to be printed by a cylinder flatbed machine.

While, by the historical accident of Bensley's interest in technical development, the cylinder flatbed machine was first applied to book printing, an obvious attraction at that time was for the printing of newspapers. A number of newspaper proprietors were invited to see Koenig's machine, among them John Walter II, of *The Times*, who had been approached earlier to join the Bensley partnership. He had declined, probably because he could see little future in an adaptation of the old screw and platen press. He had previously subsidized Thomas Martyn in the production of a new press, 'by which manual labour shall be rendered nearly unnecessary', but Martyn's so-called 'self-acting press' had proved a failure and had cost Walter some £1,500 before he brought the experiments to an end.

James Perry, of the *Morning Chronicle*, was not impressed by Koenig's new cylinder machine, but Walter, seeing its advantages, asked Koenig if he could build a double machine—that is, one which could be fed at both ends. Keonig was able to reply that he could, as he had already envisaged the possibility in his 1811 specification. Walter accordingly ordered two steam-driven, double machines for *The Times*, to be ready in twelve months, with the proviso that none were to be sold during the life of the patent within ten miles of the City of London.

Koenig's patent of 23 June 1813 contained improvements on that of 1811 and served as the basis of the double machine (Plate XXXV). For this a second cylinder was added by which the return movement of the bed was made productive. While the printing cylinders were divided into three parts as before, each being covered with cloth with points attached, the 'friskets' were abolished in favour of endless tapes conducted over rolls. The inking system underwent modification to meet the demands of double printing. The inking rollers were set transversely across the forme with their axles meeting on one side. In the patent the inking rollers were still described as covered in skin, but Koenig learned of the superiority of composition rollers during the year, otherwise *The Times* machine could not have worked as effectively as it did.

Conflicting claims are made as to the invention of composition rollers, the basic ingredients of which were eventually glue and molasses, plus a small amount of carbonate of soda, before the introduction of rubber and synthetic materials. According to Hansard, printers' composition ink-balls derived from those used in the potteries, and he supports the view that Benjamin Foster, of Weybridge, first applied composition to letterpress printing. A rival to Foster was Robert Harrild, a London printer, who, it is claimed, introduced the composition roller and ball in 1810. He said that the composition was the chance discovery of a man called Edward Dyas; but in 1819 was advertising himself as the agent for the 'improved composition' and inking apparatus of Augustus Applegath, who succeeded Koenig as *The Times* printing-machine specialist. Harrild sold Applegath rollers for his machines and, in turn, Applegath supplied Harrild with inking apparatus on agency terms. So successful was roller manufacture that in 1832 Harrild gave up printing in favour of selling printers' supplies. The first composition rollers were by no means perfect as they still retained a ridge which hindered printing, since they were moulded round a spindle and allowed to set. But by 1842 Harrild triumphantly succeeded in producing a roller without a seam and began to sell moulds for the more advanced form of manufacture.

Whoever originated composition, Bryan Donkin, who invented the 'polygonal' printing machine, had the idea of using it to make rollers, and uses the phrase in his patent of 1813. '. . . or by means of a metal cylinder covered with canvas and coated with a composition of treacle and glue'. Koenig, in his patent of the next year, refers to 'a composition such as has been used for some years past, in order to cover printers' balls, instead of skins, viz. glue and treacle'. He describes two methods of making the roller—either the more primitive way of casting the composition on a cloth and wrapping it round the roller or the more advanced way of casting upon a roller in a mould. Koenig obtained his rollers from Harrild.

To avoid difficulties with his hand pressmen, Walter installed the first Koenig machine in secrecy in a building adjoining *The Times* office, ready to print the issue of 29 November 1814. Although rumours were circulating in the press room, no action was taken by the suspicious pressmen, who had been told to wait for expected news from the Continent. At six o'clock in the morning Walter entered the press room and astonished the men by telling them that the newspaper was already printed by steam; that if they attempted violence there was a force ready to repress it; but if they were peaceable their wages would be continued until similar employment could be procured. An editorial in this historic issue of *The Times* read as follows:

'Our Journal of this day presents to the public the practical result of the greatest improvement connected with printing, since the discovery of the art itself. The reader of this paragraph now holds in his hand one of the many thousand impressions of *The Times* newspaper, which were taken off last night by a mechanical apparatus. A system of machinery almost organic has been devised and arranged, which, while it relieves the human frame of its most laborious efforts in printing, far exceeds all human powers in rapidity and dispatch. That the magnitude of the invention may be justly appreciated by its effects, we shall inform the public, that after the letters are placed by the compositors, and enclosed in what is called the form, little more remains for man to do, than to attend upon, and watch this unconcious agent in its operations. The machine is then merely supplied with paper: itself places the form, inks it, adjusts the paper to the form newly inked, stamps the sheet, and gives it forth to the hands of the attendant, at the same time withdrawing the form for a fresh coat of ink, which itself again distributes, to meet the ensuing sheet now advancing for impression; and the whole of these complicated acts is performed with such a velocity and simultaneousness of movement, that no less than eleven hundred sheets are impressed in one hour.

'That the completion of an invention of this kind, not the effect of chance, but the result of mechanical combinations methodically arranged in the mind of the artist, should be attended with many obstructions and delays, may be readily admitted. Our share in this event has, indeed, only been in the application of the discovery, under an agreement with the Patentees, to our own particular business; yet few can conceive—even with this limited interest—the various disappointments and deep anxiety to which we have for a long course of time been subjected.

'Of the person who made this discovery, we have but little to add. Sir CHRISTOPHER WREN's noblest monument is to be found in the building which he erected; so is the best tribute of praise, which we are capable of offering to the inventor of the Printing Machine, comprised in the preceding description, which we have feebly sketched, of the powers and utility of his invention. It must suffice to say farther, that he is a Saxon by birth; that his name is KOENIG; and that the invention has been executed under the direction of his friend and countryman BAUER.'

Nothing could be clearer than Walter's statements, yet after his death much was written in an attempt to deny Koenig and Bauer the honour of producing the first cylinder machine. No sooner had *The Times* machine become known than the denigration began, and Walter therefore allowed Koenig to use the editorial column of *The Times*, for 8 December 1814, to give an account of the origin and progress of his invention as 'a confused statement having appeared in several newspapers, and insinuations thrown out that the Editor of *The Times* had not bestowed the merit of the invention on the rightful owner' he hoped he would not be thought assuming if he published the facts.

Koenig makes it clear that he had to come to England because of its system of patents, which did not exist on the Continent. He then tells the story of his meeting with Bensley, who, with Woodfall and Taylor, supported him in his experiments. He benefited by the assistance of his friend Bauer, 'who, by the judgment and precision with which he executed my plans, has greatly contributed to their success'. The first machine is described, and some of the work printed on it. He continues: 'The machines now printing *The Times* and *Mail* are upon the same principle as that just mentioned; but they have been contrived for the particular purpose of a newspaper of extensive circulation, where *expedition* is the great object.'

A multiple machine had often been discussed between the two partners and Walter, and so Koenig turned his attention to devising what he called an eightfold machine, in which bed and forme would move together on a horizontal circular course, above which eight printing and inking apparatuses were to be situated. The cylinders envisaged were not strictly 'cylindrical' but were rather shortened cones, on each of which was to be a printing plate covered with cloth, upon which a frisket was fastened to prevent the sheet from moving. During one rotation of the forme each cone was to make an impression, and during the rest of the time being available for feeding with paper and taking off.

No opportunity was given to the partners to build this machine while in England as the double machine was found to be sufficient for *The Times* circulation during that period. It was not until 1823 (when they were back in Germany) that Walter thought the time had arrived to 'think of the roundabouts' (as he called them in a letter to Koenig and Bauer), although he did not pursue the matter. When he accepted the offer of Augustus Applegath to build such a machine he gave the partners notice in a friendly way, with the assurance that he would not have made use of the assistance of strangers, as he put it, if they had still been established in London.

Koenig's last English patent, of 1814, was the basis of an improved cylinder machine and of a perfecting machine, that is one which would print on both sides of a sheet of paper (Plate XXXVI). The perfecting machine was a combination of two in one, in which the forme, printing cylinder and inking device were duplicated but which had a single feeding apparatus in the shape of an endless web on which the sheet of paper was fed. A registering apparatus was fixed between the two printing cylinders, which were covered only partially to the size of a sheet so that the forme could return freely under the uncovered portion. The paper was carried between two rows of tapes round the first cylinder, to be printed on one side, and was then taken off the cylinder, laid on the register device, which sustained it until it arrived in a vertical position over the second cylinder, to be moved around it and printed on the second side. The sheet was turned by the use of an S-shaped course, and after being printed on both sides was conducted to a board in the middle of the machine. The first machine of this sort was finished in February 1816, and was installed in Bensley's office, where, steam-driven, it was used for book printing. It produced 900 to 1,000 perfected sheets an hour. The second edition of

Dr. J. Elliotson's translation of Blumenbach's *Institutions of Physiology* was, in consequence, the first complete book to be printed by a machine.

The patent described improvements in the printing and inking apparatus, the ink-box being placed in immediate contact with the distributing cylinders and 'the action of the frame of the lower inking cylinders is simplified by giving a separate frame to each cylinder', with a 'common centre for both'.

As well as the perfecting machine, Koenig and Bauer produced the improved single machine, with a feeder and register device, and which was the forerunner of the 'drum cylinder' type of machine. It printed from formes measuring 36×25 inches at a rate of 900 to 1,000 an hour, and was installed by Richard Taylor for book printing. Later, when Taylor obtained the contract for printing the *Weekly Dispatch*, Koenig and Bauer altered it to a double machine, although by this time they were back in Germany and had to send the parts to London.

The Times also required alterations to speed up printing of between 1,500 and 2,000 impressions an hour, and improvements were made to its machines. On the earlier machine the cylinder was arrested for each single sheet; on the improved machine the cylinder was in continual motion, there being only one printing area covered in felt, and a feeder over each cylinder, which remained motionless as long as the feeding required and then delivered the sheets to endless tapes. It was thus the first 'two revolution' printing machine—one which makes an impression at every second turn of the cylinder.

Koenig and Bauer by 1816 naturally wished to manufacture for the printing trade in general but Bensley, principal shareholder in the company, refused to allow this. He had become the major shareholder in 1813 when Woodfall ceded his share to him. Despite the appeals of Koenig and Bauer and the other shareholder, Taylor, Bensley maintained that machines should be for his benefit only. The situation became intolerable for Koenig and Bauer and they decided to return to Germany to manufacture there. They signed a deed of partnership on 9 August 1817, and the Koenig and Bauer factory was established at Oberzell, near Würzburg.

Only a very few printers in England were ready for the cylinder printing machine—mostly in the newspaper and periodical field—when Koenig left for Germany, and the printing trade of his own homeland was completely unprepared for this advance, as he found initially to his disadvantage. Johann Spener and Georg Decker had written to him in 1815 when he was still in England, asking for particulars of his machine but they found the price too great. Decker therefore asked him to send him a *Stanhope* press, and this was the first ever introduced into Germany. Some idea of the economics of printing presses can be obtained when it is realized that the *Stanhope* cost £95, whereas the simplest Koenig machine, the single-cylinder, non-registering machine, cost £900. A double machine cost £1,400 and if it had the registering device it was £2,000, and added to these prices were £250, £350 and £500 a year for each of these machines so long as the patent lasted, unless an agreed sum was to be paid down at once.

Apart from the cost factor there was also the human element to be taken into account. Pressmen did not like the new machines. According to the famous New York printer Theodore Low de Vinne, whose experiences went back to 1843, they called them 'type smashers', and despite their original use, in Bensley's and Taylor's offices, for bookwork the tendency was to use them for cheaper newspaper and periodical printing only.

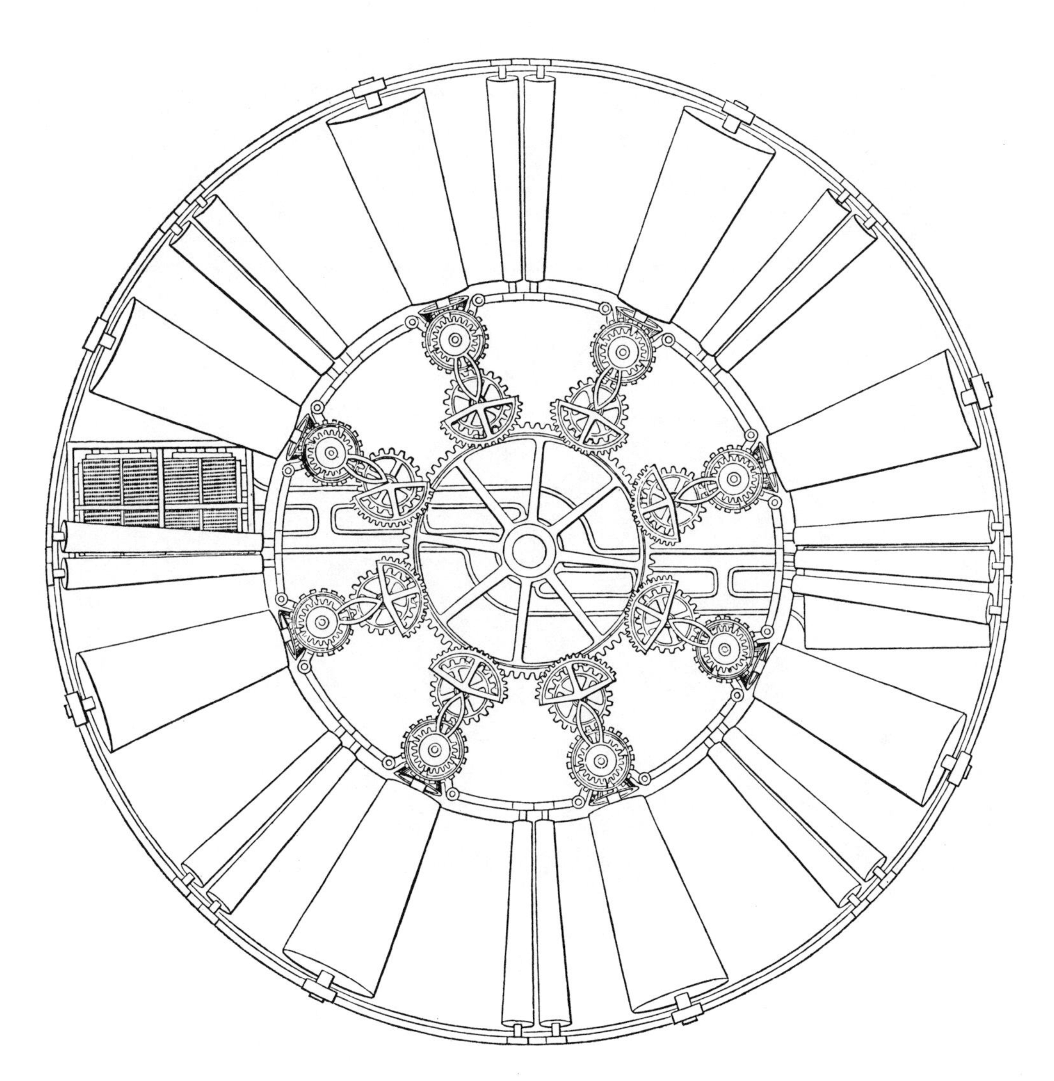

47. Diagram of the projected 'roundabout' or multiple machine designed by Koenig for John Walter. (See page 109)

48. One of thirty Adams bed and platen presses at work in the printing office of Harper Brothers, New York, 1855

8

THE BED AND PLATEN PRESS

The cylinder printing machine and the application of steam-power were not at first widely adopted by the printing trade for a number of reasons. The cylinder required more careful handling than the hand press, and by men with some mechanical skill. Among the hand pressmen there was a degree of prejudice against the machine, and for many years the pressmen and the 'machine managers' maintained separate trade societies.

Apart from craft prejudice and the difficulty of obtaining skilled machine managers there were sound economic reasons why the cylinder machine was not universally adopted. Until it was improved it could not print the better class of work, and was restricted to newspapers and periodicals. Its productivity was also too great for the comparatively small runs required by many printers.

At the same time, even the most traditionally minded printer could not fail to recognize the advantages of the new inking rollers over the smelly ink-ball, and the benefits of power operation if it could be applied to a simpler machine. There arose, therefore, a demand for something in between the hand press and the cylinder machine, and men such as Joel Northrup (in the United States) and Johann Deisler (in Germany) tried to meet this by increasing the number of printing units on a hand press. These efforts were not successful and the answer was finally found in the so-called 'bed and platen' press, which was based on the same principles as the hand press, but could be power-driven. Hand pressmen could carry out the time-honoured techniques of packing the tympan, fitting the paper to it and lowering the frisket. Pressure was downward and applied by a platen, but there was no need to pull a bar. High-quality bookwork could thus be produced at a faster rate than on the hand press.

The man who first came on the idea of the bed and platen press was Daniel Treadwell, who returned to the United States from England in September 1820, after the failure of his treadle press. His stay did have one effect—it had enabled him to study the cylinder press in action and to calculate the advantages of steam-power. When he arrived back in Boston he wanted to use this new-found knowledge to manufacture his own cylinder machines.

Although he called on his fellow townsman, Phineas Dow, a machinist, to assist him, because of the lack of adequate machine tools it was found impossible to construct an efficient machine. A press of a kind was certainly made, but as no steam-engine was available it was driven by horse-power. At first the press was built mainly of wood, metal gradually being introduced. It had a horizontal and reciprocating bed which was put into motion by a strong vertical iron shaft. By the use of cams and pulleys the bed was brought under the platen and the platen brought down—the actual pressure being supplied by a toggle joint operated by a fly-wheel. The inking rollers were supported in a carriage, which took them back and forth over the type after each impression. At the rear of the press was a circular, horizontal, intermittently rotating ink table, which made one-eighth of a revolution for each impression, the inking rollers passing over it at each of these. This device was later adopted by the pioneer of 'jobbing platens'. Treadwell's tympan and frisket were much the same as those on a hand press. While the machine was crude and had many drawbacks, it is said to have printed at the rate of 500 to 600 sheets an hour. Treadwell sold the rights to build and use his press to others, and also built two presses back-to-back, which could be driven by a common shaft. Steam-power eventually replaced horse-power.

49. The hand-operated wooden frame Adams bed and platen press

The virtues of the bed and platen press became apparent in both Britain and the United States. The English-born Jonas Booth, who emigrated to New York in 1822, claimed as the first in America to apply steam-power soon after his arrival, produced a press similar to that of Treadwell, but which had a frisket at each end so that a sheet could be placed on one frame while another sheet received the impression from a single forme. Booth patented his press in 1829, but did not sell many models. A rival claimant was Daniel Fanshaw.

By 1830 the brothers Isaac and Seth Adams, who had been manufacturing iron hand presses, went over to the 'bed and platen'. Their basic conception remained popular for more than fifty years. The press had a stationary platen and a bed which moved up and down to press the forme against it. The paper was fed from above the end of the press, being taken to the point of impression by a frisket. When it reached this point the bed rose, made the impression and fell again, the printed sheet being taken to the other end by tapes. Originally the press was made principally of wood, and was driven by man-power, but the Adams brothers later substituted iron parts and steam-driven equipment.

Not very kindly, a writer in the *Typographic Messenger* described the first *Adams* press as a 'clumsy hard-wood clap-trap', and attributed improvements to a Mr. Austin, whom Isaac Adams threatened with a suit for infringement. Austin appealed to the 'English books on printing machines', and claimed an equal right to the principles embraced. Presumably he was able to point to the *Napier* and similar presses—which will be described. Being frustrated in one direction, Adams improved his press, perfecting the register and adding an automatic sheet-pile (and patenting the whole mechanism) until Austin was put out of business. Others were interested in this type of machine and during 1839–40 Hoe & Co. built two bed and platen power presses for Samuel Fairlamb from his designs, to compete with the *Adams* press. Both bed and platen rose and fell so that they balanced each other. Two friskets were alternately brought in between bed and platen by the motion of the inking-roller carriage. According to Stephen Tucker, later a Hoe partner, the *Fairlamb* was a simple machine but deficient in speed and power compared with

50. The Adams bed and platen press as built by Hoe

the *Adams* press. So between 1846–7 Hoe built a press similar to that of Adams, who claimed infringement of patent. After years of dispute, in April 1859 Hoe's purchased the entire Adams establishment and unexpired patents, and began to make the *Adams* bed and platen machine with various improvements. The firm also made a 'bed and platen job printing press', based on the *Adams*, which, perhaps, should be dealt with in the chapter on jobbing platens, but it is convenient to do so here as it derives from the other press.

Meant for small work, such as cards, the press was worked by a treadle, but the fly-wheel was adapted to receive a belt, so that steam-power could be employed. The sheet of paper was placed on adjustable guides on to an inclined platen and was lifted to the bed by means of a cam on the main shaft. The ink was taken from a fountain by a ductor roller to a metal cylinder on which a vibrating roller was constantly traversing.

The *Adams* bed and platen press had a considerable influence on the American printing trade. It eventually reconciled the hand pressmen to power-driven machines and enabled them to produce first-class work at speeds of from 500 to 1,000 impressions an hour, of which they were proud. It must have been the only press to have given its name to a trade union—the Adams and Cylinder Press Printers' Association, originally in 1865 exclusively for Adams power pressmen in New York, but later extended to cylinder pressmen. A venerable *Adams* press was still being operated at the Riverside Press, Cambridge, Massachusetts, as late as 1938.

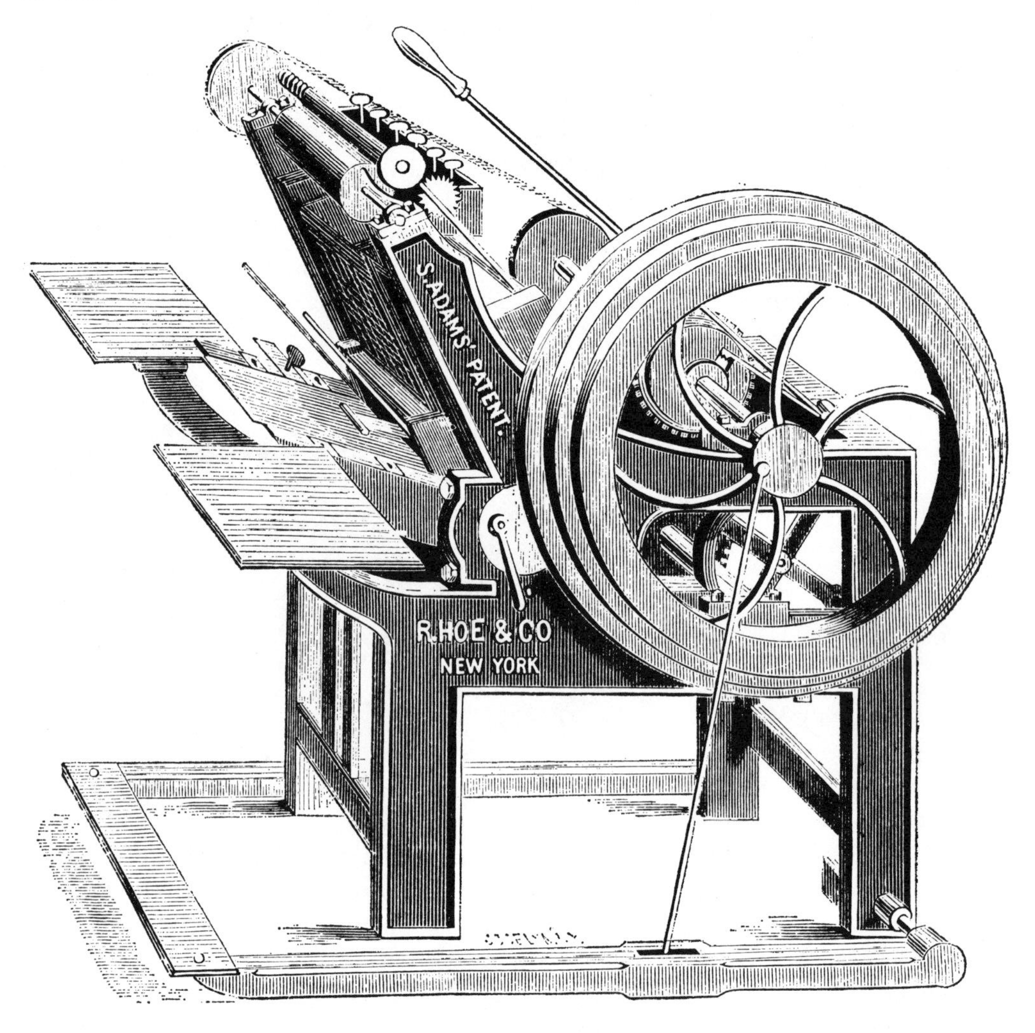

51. The Hoe Adams bed and platen job printing press. (See page 115)

In England the 'improved' double-feeder bed and platen made by the firm of Napier, and originally conceived in 1830, became the most popular press of the kind in the trade. The platen was raised and depressed by a pair of connecting rods—one on either side of the frame. William Spottiswoode, who had already installed an Applegath cylinder machine, introduced a *Napier* platen to work beside it. Stephen Austin, of Hertford, put one in at the same time as his *Main* cylinder machine, for which he had had to travel to London to find an operative. He put the platen in to produce good-class bookwork (for which he became famous) and kept the cylinder for his newspaper.

A *Napier* double-feeder platen continued working at the Oxford University Press until well into the twentieth century. In 1930 Charles W. Jacobi, the famous manager of the Chiswick Press, wrote of it in the *Penrose Annual*: 'An interesting relic of the past is one old-fashioned double platen machine, many of which class were introduced about 1837. About that time they were considered to give an ideal impression—that is from the flat, as compared with that of cylindrical printing, and the most suitable for bible and similar work. This relic is still in use and employed only for special work.'

The very originators of the cylinder press, Koenig and Bauer, were so impressed by the bed and platen machine that they built one in Germany on Napier's system. In its turn, during the 1850s, the firm of Napier replaced its cylinder business for that of 'bed and platens'. The Bank of England had tried a *Napier* double cylinder (sold for £470 in 1852) for printing notes of the highest standard, but within two years the Bank authorities had changed their minds in favour of the double bed and platen machine, patented by J. M. Napier in July 1853. Two machines were sold to the Bank at the very high price of £2,790. They were capable of producing about 1,500 bank-notes an hour. A special characteristic of the machine was the fact that the printing plate was inked four times by each of the rollers—the equivalent of twenty hand inkings. The *Napier* machines were not superseded at the Bank until 1881.

Napier & Son then concentrated on selling bed and platen machines to firms which produced bank-notes, postage stamps and similar work—such as De la Rue. Eyre and Spottiswoode at one time had twenty-two machines to print Bibles, and the University Presses followed suit. Overseas governments, including those of Germany, Italy and Japan, bought machines for printing currency notes and the Imperial Russian government proved a particularly good customer, buying ten machines at the special rate of £400 each in 1865. When in 1887 the Government of New South Wales, Australia, began printing postage stamps by letterpress, a Napier double platen was imported from England for the purpose. A similar press was also used by Enschedé en Zonen, of Haarlem, for printing Dutch postage stamps up to 1914.

Another bed and platen machine was that of Carl August Holm, originally of Stockholm, who, under the patronage of Count Rosen, devised his *Scandinavian* press in London in 1841. In this press the single platen was raised and depressed by rods on each side of the frame. Originally the manufacture was undertaken by Braithwaite, Milner & Co., of Bath Place, New Road, London, but there was some criticism of the noise of working. The manufacture was therefore transferred to Hopkinson and Cope, the manufacturers of the *Albion* press, who stated, in their 1862 catalogue, that they had lessened the noise and had strengthened the press. With an eye on the hand pressmen they maintained that all the means of 'making-ready' were the same as at the common hand press, and stressed the superiority of the press for printing from wood cuts. At that time no printer would have chanced printing direct from wood cuts on a cylinder press, which might damage them. The *Scandinavian* press was made in two sizes—Super Royal (28 × 21 inches) and Double Crown (31 × 22 inches)—the smaller size being capable of being worked by hand or steam, the hand apparatus costing £4 10*s*. extra.

T. H. Dodge. Sheet 1. 2 Sheets.

Printing Press.

No. 8521. Patented Nov. 18. 1851.

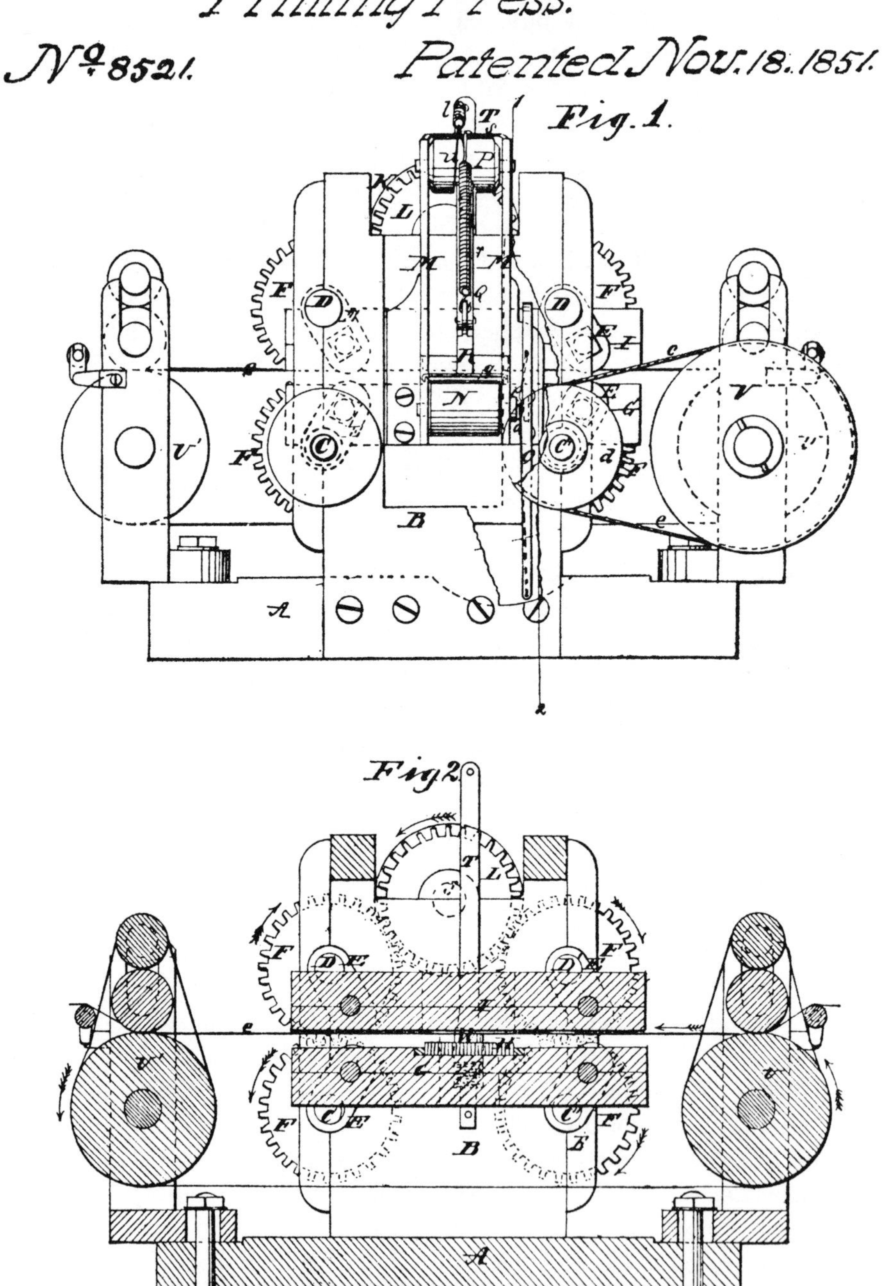

52. Diagrams of Dodge's reel-fed bed and platen press. (See page 119)

However, for larger sizes the Hopkinson and Cope 'Double Platten' or Flat surface press was recommended. This had been developed in 1830 and it would have been more accurate to call it a double-feeder press in the same way as the *Napier*. The platen was placed across the centre of the side frames, supported by a powerful beam (reminiscent of that on the *Columbian* press), one end of which was held firm in a bearing, while the other worked freely up and down between guides. There was a table carrying a forme at each end of the press, moving alternately under the platen. Speeds of from 1,200 to 1,600 impressions an hour were claimed.

The Double Demy size ($35 \times 22\frac{1}{2}$ inches) cost £340. If fitted with the 'improved Mouse Roller Distributing apparatus' at both ends a further charge of £30 was made.

Other manufacturers saw the attractions of the 'bed and platen' press, including the increasingly active firm of Harrild & Son, which advertised in 1860 its 'improved double-platen machine for book-work', in Double Demy size ($35 \times 22\frac{1}{2}$ inches) at £390; and its 'patent Scandinavian platen machine', which cost £139 in the Super Royal size (28×21 inches) and £172 in the Double Crown size (31×22 inches).

In Paris, Marinoni constructed bed and platen presses for the Bank of France, which used them to print postage stamps and bank-notes. They printed at the rate of 900 impressions an hour. This was in the 1870s, but an earlier, unsuccessful, attempt had been made in 1855 to build a machine by Victor Derniame.

Thomas H. Dodge, of Nashua, New Hampshire, had the notion that a bed and platen press could be fed from a reel of paper. While looking at a railway engine it also occurred to him that the parallel rod connecting the driving wheels supplied the kind of motion he sought in a press. Accordingly he built a press, for which he received a patent in 1851. The bed and platen moved in the same direction longitudinally, and an impression was made at every revolution of the cranks. Paper was fed from the reel at the same speed as the motion of the bed and platen. However, printing was on one side of the paper only and two presses, one under the other, were required if printing on both sides was necessary. What the ultimate destiny of this press might have been in the letterpress printing field is not known, as the patent rights were bought by a manufacturer of cotton bags and it ended its days as a textile printing press.

One other inventor, F. Tilgmann, of Helsingfors, Finland, applied reel-feeding to the bed and platen press. The machine, called the *Miu*, after the inventor's deceased wife, is described in Frederick Wilson's *Typographic Printing Machines* (Wyman & Sons, London, 1879). Unlike other bed and platen machines the forme was forced up to a stationary platen, rather in the manner of the *Adams* press. The roll of paper passed down in a perpendicular direction, then under the platen where it received the impression. It then ran between a roller and pulleys, and at this point could, if desired, be cut into separate sheets. In the case of wallpaper, for which the machine was particularly adapted, instead of being cut, the paper was conducted over another roller and rewound preparatory to the next working.

The machine was adapted for the printing of linen, cloth or leather, and was usually run at the rate of 800 impressions an hour, but, said Wilson, this could be increased to as many as 1,500, although at this speed both the register and the

D. Napier & Son's Patent Double Platen Printing Machine.

Sizes at present Manufactured.

No.	Forme of matter in inches.			Impressions per hour.
1	43	by	27	1300
2	40	"	27	1300
3	36	"	21½	1400
4	28¾	"	18¾	1600
5	21½	"	16	2400

53. Napier's double bed and platen press, with details of size and performance

distribution of the ink were liable to become defective. While the machine could be driven by hand this was inadvisable as the required regularity of motion could not be assured.

The bed and platen press, so long as it performed a useful purpose, continued to be used alongside the cylinder machine, but it was not capable of very great development, and fell into disuse as progressive improvements were made to the cylinder. Wilson's remarks (1879) are interesting in this respect: 'Platen machines have long enjoyed the reputation of being capable of producing the finest work that can be done on a machine; whether that reputation will be maintained in face of the improvements that are constantly being introduced into cylindrical machines is, however, extremely problematical.' Nevertheless, for many years printers used different types of press and machine at the same time, often in one workshop. However, large printers began to separate the press from the machine. Between 1830 and 1832 the Spottiswoode brothers, in London, bought an Applegath cylinder and, for the first time, there was a division into press and machine rooms, which, as the following illustrations of the Brockhaus Leipzig establishment in 1866 indicate, became an established practice.

54. A Brockhaus press room of 1866. On the left are two rows of German-built Columbian presses, identified by griffin counterweights (see page 66). They stand on wooden 'tees' derived from the Stanhope press. The presses on the far right are too indistinct to identify, but that in the foreground (and those behind it) are probably either Hoffman presses or Säulenpressen, copied from the Cogger. (See page 72)

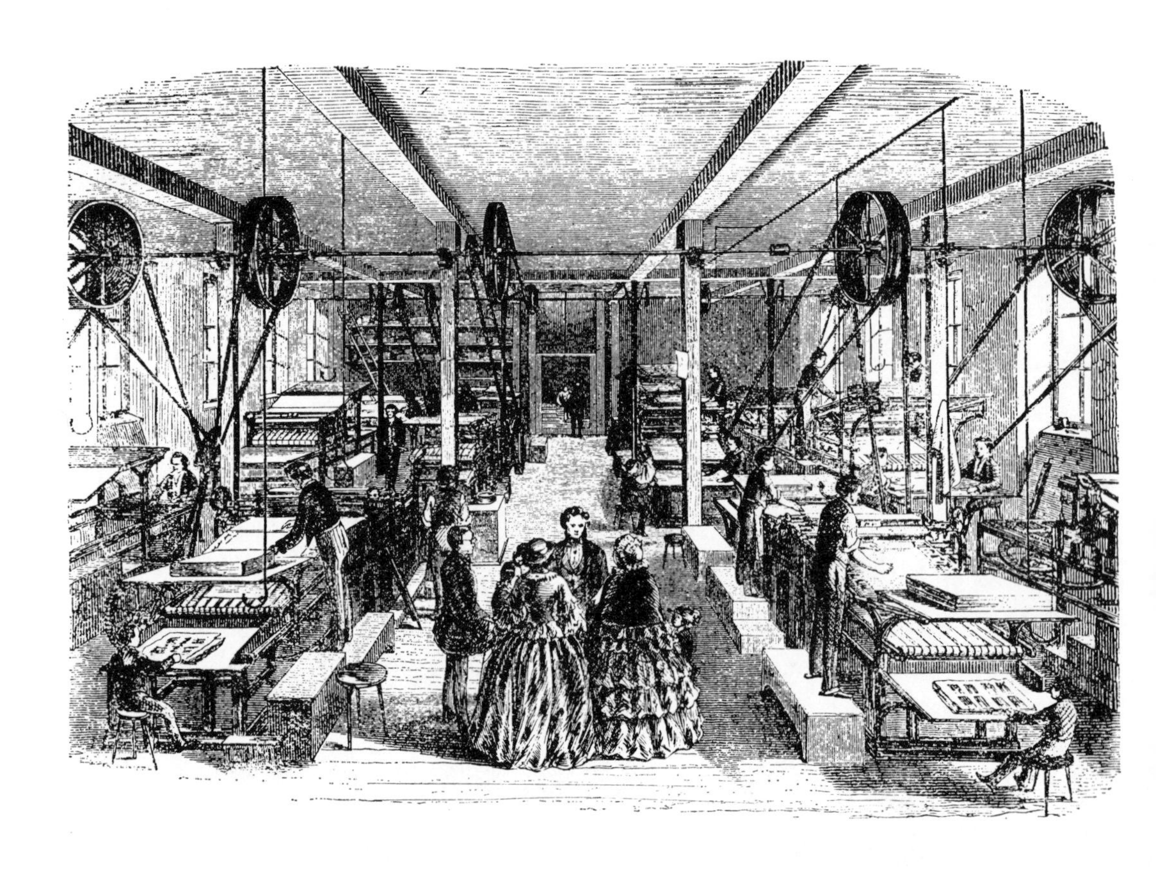

55. A Brockhaus machine room of 1866. The machines were probably made by Koenig & Bauer. That on the far right has the characteristic Bauer 'sun-dial' movement. (See page 139)

9

THE FIRST STAGES OF THE CYLINDER MACHINE

ALTHOUGH THERE HAS been a tendency to equate the emergence of the cylinder printing machine with the application of steam-power, the two developments were not necessarily interrelated. Power, other than that supplied by a human being, could be and was applied to other presses besides those constructed on the cylinder principle, and some cylinder machines were worked by hand. Efficient steam-engines were not universally available for another thirty years after Koenig's invention. In 1820 there were only eight steam-presses in the whole of London, nearly all being used by newspapers, except for those of Strahan, the King's Printer. As late as 1851 the Printing Machine Managers Trade Society had a membership of only 130, indicating that in London, the main centre of printing, there could not have been many machines at work.

Charles Manby Smith, whose memoirs present such a vivid picture of early nineteenth-century printing, in his *Curiosities of London Life* (1853) has an article 'The Irish Machine' which is about Terence O'Donough, who successfully replaced a steam-engine to operate a printing machine in London; in 1863 the *Greenock Telegraph* was printed on a single-cylinder machine operated by a 'squad of navvies', and Will Ransom recalled that in the eighties at a newspaper called *Iron Ore,* printed in Northern Michigan, when the motor power failed, as it did frequently, he was sent out to recruit off-shift miners to work the press by means of a crank attached to the fly-wheel.

Nevertheless, in the long run, it was the combination of the cylinder machine and steam-power which began the transformation of the printing trade, which had remained substantially unchanged for four centuries. Koenig and Bauer had, in effect, laid down the basic elements of the cylinder machine in the early part of the century, and while many improvements took place, the principles did not alter for more than a hundred years. The four basic types were the stop-cylinder, in which the cylinder stops after making the impression and remains stationary during the return of the bed carrying the forme; the two-revolution, so called because an impression is made at every second turn of the cylinder; continuous revolution where the large drum-cylinder is twice the length of the bed, the impression part alone being covered with a blanket, the difference in diameter between this part and the remainder being sufficient to allow the bed to return while the cylinder completes its revolution; and the oscillating or reversing cylinder where the cylinder which accompanies the bed in its forward movement is then raised and swung or rocked back to its original position.

56. First successful drum cylinder flatbed press (Hoe, 1830). (See page 123)

There was also a still earlier principle, conveniently known as the travelling cylinder, which, to a certain degree, reverted to the idea of rolling a cylinder over an inked forme, visualized as far back as the seventeenth century (see the opening of Chapter 7). A machine embracing this idea was invented by Edwin Norris, of Walworth, in 1835. A sheet was printed each time a pressing cylinder, in a carriage, passed over a type forme fixed in the centre of the press. These machines were manufactured by Carr and Smith of Belper, in Derbyshire, and hence gained the name of *Belper*. Among the newspapers to install a machine was the *Inverness Courier*, but since it was capable of producing only some 500 impressions an hour it had been superseded by the mid-nineteenth century.

In time, gas-engines replaced steam-engines, and finally electric power was used to drive the machines. At an early stage it was realized that if a sheet could be printed from a single cylinder then two sheets could be printed from two cylinders, and there thus developed the idea of multi-cylinder presses. Equally, there was the possibility of using say two formes, and two beds, and one cylinder to print two colours on one side of the sheet, and then onwards to two or more cylinder units in tandem for printing two or more colours. The perfecting machine already used two cylinders for printing on both sides of the sheet, but it, too, could be developed for printing in two or more colours.

An example of the desire to utilize all the elements of a printing machine to increase productivity is shown in the patent taken out by John Gedge, of Wellington Street, Strand, in 1852. He wrote: 'I intend to alter the arrangement of the printing machines in general use, (more particularly those used in newspaper or other large printing establishments where great celerity is required,) and to make my machine consist of two "pairs" of cylinders, instead of two cylinders; one pair being employed in printing one side of a sheet of paper, and the other pair perfecting or printing the other side. I employ two pairs of cylinders in order that I may be enabled to divide each forme (printing form) into two parts, making, when ready for use, *four* small forms. The sheet of paper is passed by the tapes, &c. under the first and second cylinders, and one side is printed; the sheet is then carried over by guides to the other pair of cylinders, and the opposite side of the paper is by them and the two forms under them printed; thus, it receives its impression from four distinct forms by the action of four distinct cylinders. The object sought to be attained by the Invention is, an increased rate of production without increasing the speed of working the press or machine.'

The Gedge notion of multiplication of elements to increase production was followed for some time until it was realized that in the long run the machines would reach a point where they would become unmanageable. This was when the idea of reel-fed rotary presses began to take hold.

But all these developments in the source of power and in the capacity of the machine took time to achieve and, in fact, the first cylinder machine to follow those of Koenig and Bauer was a very humble affair, of the hand-operated variety. Invented by William Rutt in 1819, and built by Baisler and Napier, who had been responsible for Treadwell's press, it was based on Koenig's first machine. The single cylinder made one-third of a revolution and then came to rest to allow the type bed to return. The bed was driven continuously to and fro along the press by means of a mangle wheel or crank and rod. Composition rollers were by now in use, but Napier had not yet invented grippers so that the sheet of paper was carried by two sets of silk bands to the point of impression, after which it was discharged by two others on to a set of four ribbons, from which it was removed by hand. Hansard provides an illustration of the machine, from which it is seen that it required three men to operate it, and it must have therefore been very time-consuming.

Following Koenig and Bauer's departure for Germany, users of their presses, notably Walter and Bensley, sought the services of two other men to improve them, and from this fact there arose further denigration of the Germans. The men in question, a printer Augustus Applegath, already mentioned, and his mechanic brother-in-law, Edward Cowper, no doubt did improve what were, after all, pioneer machines, but it is not certain, for example, that they were responsible for increasing the speed of *The Times* machine from 1,000 to 1,800 copies an hour, as

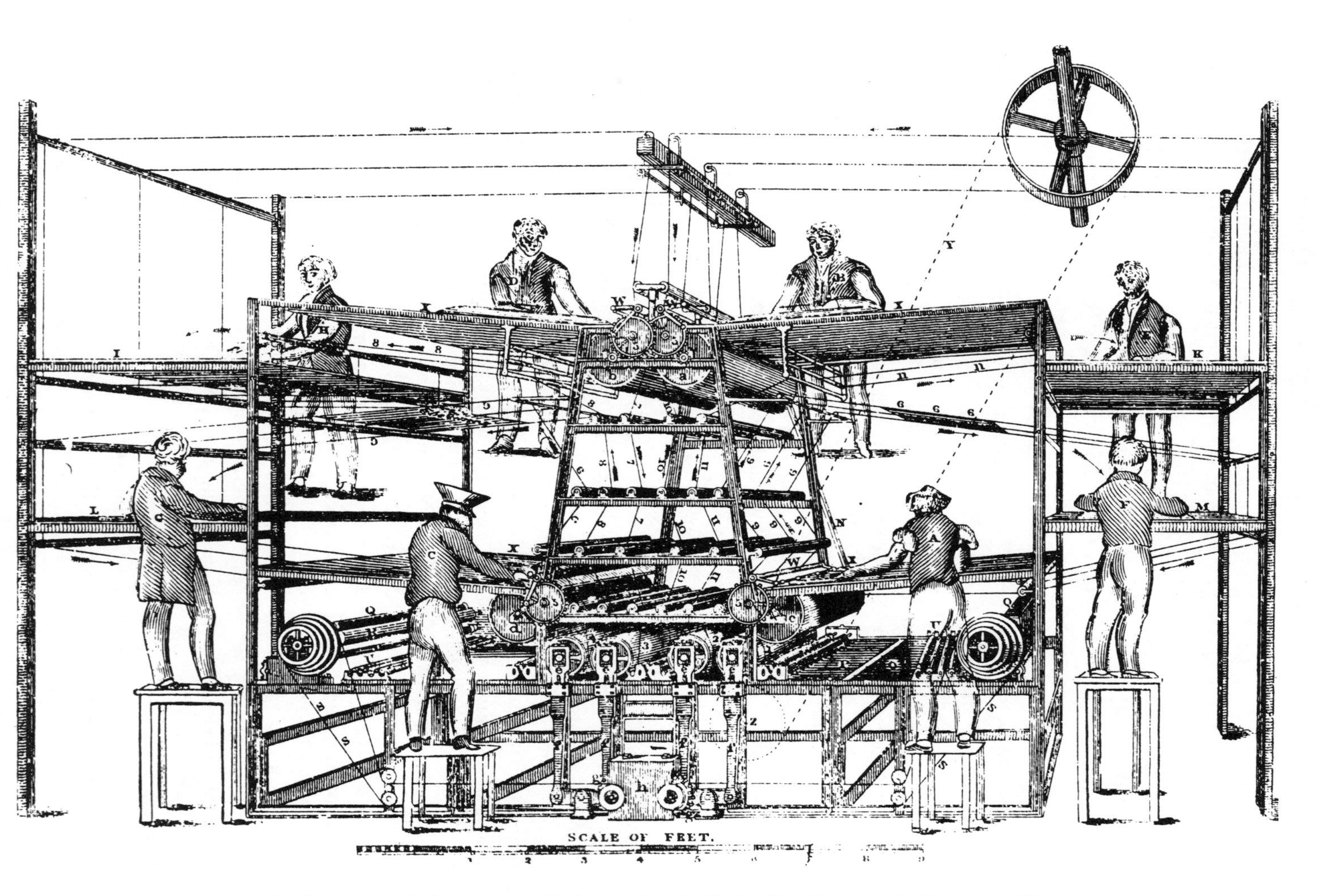

57. The Applegath four-feeder cylinder machine built for *The Times*, 1828. (See page 129)

has been stated. This improvement was in hand before Koenig and Bauer left England and was one of the last jobs undertaken by Bauer in London.

Applegath and Cowper first came into prominence when the Bank of England became concerned at the extent of bank-note forgeries. The two men worked in partnership to find ways of beating the forgers. In 1816 Cowper patented a method of printing 'paper for paper-hanging and other purposes' in one or more colours from curved or bent stereotype plates made by the plaster of Paris process and heating. The plates could be fixed to a cylinder, inked by composition rollers and printed from the rotary method. The two men were granted money to experiment on note-printing with Cowper's device, but, in the long run, the Bank did not continue with its support.

For some reason, neither Cowper nor Applegath followed up this early idea of rotary printing, possibly because there was no demand at that point. While working for the Bank they had also been employed by Bensley and Walter to supervise Koenig's machines, having already been engaged by Bensley to improve the perfecting machine, particularly the inking mechanism. In these improvements Applegath and Cowper made use of their own patented devices, and removed some of the imperfections. This does not mean that Koenig and Bauer could not have done this themselves, but they were far away in Germany.

In January 1818 Cowper patented his ink-distributing table, which was attached to the forme, and indentations at its sides gave an endwise motion to two distributing rollers in a movable carriage held on four bearings, and with two small friction pulleys attached. The ink was conveyed by a vibrating roller which was alternately in contact with the table and with a 'ductor or doctor' roller turning in an ink trough. The table and forme both passed under the inking rollers, which received ink from the table and inked the forme as it passed under them. In Cowper's specification the rollers are described as 'covered with leather, felt, composition (treacle and glue) &c.', an indication that he was still gradually working his way towards composition rollers at the time.

Another Cowper improvement concerned the method of conveying the sheet of paper from one cylinder to another in a perfecting machine by the construction of two subsidiary 'carrying drums' between the impression cylinders, on which the sheet was carried by means of two sets of endless strings, 'each composed of two or more strings kept tight by weights or springs', the printing cylinders and carrying drums being connected by means of toothed wheels.

Applegath in 1823 patented a system in which the distributing rollers or wavers moved diagonally across the forme, thus giving them an end motion. He took out a number of other patents, mostly of an experimental nature, but one, in 1824, for the rocking cylinder was adopted. The idea was to enable two cylinders to act alternately by means of a rocking motion upon the same forme, which had to travel through a short distance only before giving the second impression. This was used in a machine produced by Applegath for *The Times* in 1827.

For, after the experience gained by working on the Koenig machines, Applegath had decided, in 1821, to set up as a printing-machine manufacturer, taking premises in Duke Street, Blackfriars. It has been customary to refer to the machines he built for *The Times* and others as the work of both Applegath and Cowper, but it is by

58. The Alauzet four-feeder cylinder machine. (See page 129)

no means certain that Cowper remained in partnership with Applegath after the Bank of England episode. Cowper clearly thought he had been partially responsible, as, in a lecture before the Royal Institution on 22 February 1828, he referred to: '... *The Times* machine which was arranged by Mr Applegath upon our joint inventions.' William Savage, in his *Dictionary of the Art of Printing* (1841), refers to Applegath as the sole manufacturer of the machine. Cowper also claimed that 'our machines' were used by the most celebrated printers of London, Paris and Edinburgh. What he meant by 'our' is not clear, as while Applegath was not entirely successful he seems to have been working on his own.

Because the circulation and number of pages of *The Times* was increasing John Walter required a new and more productive printing machine, and he approached Applegath to solve his problem. A machine with the capacity of four presses was devised in 1827 by Applegath. Driven by steam, this 'four-feeder' was about 13 feet high and 14 feet long. The type bed travelled to and fro almost the whole length of the machine below four impression cylinders grouped together, and the forme was inked on each journey by four sets of composition rollers. Sheets of paper were fed from four boards. The rocking system was applied to the cylinders, which rose and fell alternately, one of each pair receiving the impression on the outward journey and the other pair on the return. The machine worked at a speed of 4,200 impressions an hour and continued to print *The Times* for another twenty years. The French manufacturers Alauzet and Marinoni did not begin to produce four-feeder cylinder machines until the one at *The Times* was outdated.

Applegath obtained the assistance of Thomas Middleton, an engineer in Southwark, to build this four-feeder, and, later, Middleton produced machines for other newspapers based on Applegath's designs. Middleton, in his 1862 catalogue, summed up the aims in printing-machine manufacture in the period before 1850, when he wrote: 'Certainty of register and high speed were the points to which the attention of engineers were most steadfastly directed. Messrs. Applegath and Cowper successfully grappled with the first by the addition of two rollers between the impression cylinders; one for reversing and the other for retaining the sheet in position between the first and second impressions. This improvement, coupled with a plane table instead of a complicated system of rollers for ink distribution, engrafted on König's machine, being the model, as far as regards first principles of construction, on which the best perfecting machines of the present day are built.'

By 1826 Applegath was in financial difficulties and he decided to sell his Duke Street workshop to William Clowes, a progressive book printer, for whom he had built a number of cylinder machines. He had also built a perfector for Bensley, to replace the old improved Koenig machine which had been damaged in a fire. The new perfector is illustrated in Hansard's *Typographia* as 'Dr. Bensley's Printing Machine'. This seems to have been a habit with writers (perhaps stimulated by Bensley himself). The *Encyclopaedia Londinensis* (1826) carries an engraving of 'Bensley's Steam Engine' (the perfector) together with a long description, without mentioning Applegath or Koenig. In the *Literary Gazette* of 26 October 1832 'Bensley's machine' is mentioned, a mistake which was perpetuated in earlier editions of the *Encyclopaedia Britannica.*

59. The *Illustrated London News* proudly described its 'steam printing machine' in its issue of 2 December 1843, with this illustration. In effect, there were two separate double-cylinder machines, one for printing each side of the paper. Each machine had two feeders and two takers-off. Manufacturer was Thomas Middleton, who had assisted Applegath in the construction of *The Times* four-feeder, which was still working at the time of this report, producing about 4,000 impressions an hour. The *I.L.N.* combination worked at the rate of 2,000 perfected impressions an hour. The next year the *I.L.N.* proprietors bought a Napier perfecting machine

Applegath gave up full-time printing-machine manufacture in 1827 and concentrated on silk-printing. However, in 1830, he patented a machine for printing on rolls of calico or silk from bent intaglio plates, and built two other specialist machines—one for printing coloured maps for the Society for the Promotion of Christian Knowledge, and the other to take a forme 7 feet long on which he printed wall charts for singing for the Tonic Sol Fa Co. He returned to newspaper-machine manufacture before 1848 at the request of *The Times*.

Once Koenig and Bauer had pointed the way to the cylinder machine it was natural for others to try their hand, but many of the projects did not get beyond the prototype stage. Hansard's *Typographia* is the source of information about early efforts, but his judgement is to be treated with caution as he had an interest in his 'own' machine. A very early effort was that patented in 1818 by Charles Brightly, a printer, and Bryan Donkin, a well-known engineer, which had two formes and a pressing cylinder. The formes were to be propelled over and under each other alternately upon an endless railway. Hansard has this to say: 'A machine to work by hand was also invented by Mr. Brightly, printer, late of Bungay in Suffolk; and made by the celebrated engineer, Mr. Donkin. The inking and pressure of this machine was, like all the forementioned, by cylinders: but the movement of the formes was different. They were brought alternately under the action of the pressure and inking cylinders, by rising and falling, and by passing over and under each other. The machine itself was a beautiful piece of finished mechanism; but I saw many defects in its operation of printing. I believe only one of these was ever made; and that has become, as Rowe Mores says of his letter-founders, a *nulli-biquarian.*'

The 'forementioned' included a number of machines, of which Hansard thought little. An invention discussed by him was the press invented by Robert Winch, of Shoe Lane, and patented in 1820. This was based on two pressing cylinders, rolling one on each side of a double inclined plane, and being driven by chains passing over a cylinder at the top of the incline. Hansard could not discover what happened to the machine, but, from a historical point of view, Winch has some importance as he put forward the idea of the travelling cylinder. Curiously enough, eleven years later, in another patent, Winch reverted to a 'cast-iron platen (in the middle of the press)', which was screwed to a movable beam which moved up and down by the action of guide rods. There was, however, to be an automatic inking apparatus. Why Winch dropped the idea of cylinders in favour of the older platen can perhaps be explained by the fact that there was not a great demand for cylinder machines and that he was responding to the demand which, as outlined in Chapter 5, resulted in the 'bed and platen press'.

Hansard dismisses the *British and Foreign* press of Samuel Cooper and William Miller (patented 1821). This had a carriage running from end to end of the press, where it was received on spiral springs, and being driven by chains round the pressing cylinder. Nevertheless, it foreshadowed the oscillating and reversing, or rocking cylinder, which Applegath was to develop.

A successful machine manufacturer was to emerge in the person of a Scot, David Napier (1785–1873), who had come to London and set up his own general engineer ing workshop in St. Giles. As noted, he had already worked for Rutt and Treadwell and later tried his hand with a self-inking *Albion.* Then in 1824 Hansard commissioned a perfecting machine from him. Known as the *Nay-Peer* it gave greatly improved register and embodied a system of 'grippers', actuated from inside the cylinders, for taking hold of the sheet as it was fed to the first cylinder, holding it while the first side was printed and then releasing it at the point when a corresponding apparatus on the perfecting cylinder received it. Moreover, the machine incorporated a rocking cylinder, but it is not known whether Napier invented this independently or copied Applegath, or, indeed, whether both men obtained the idea from Cooper and Miller. Napier made several machines for Hansard, originally to be worked by hand crank and then in 1832 to be driven by steam-power.

The *Nay-Peer* was not patented, but in 1828 Napier took out a patent for the application of a four-feeder apparatus to a machine with a single pressing cylinder. The inking apparatuses, said Napier, were the same as those adapted to his

60. The Napier Gripper machine

'Naepeer' [*sic*] machines ('which machines are well known to the public'), with the exception that they were moved by wheel and pinions from the cylinder rather than by rack and pinion.

Napier's next patent (1830) was important as it carried forward a major development—the 'two-revolution' principle which eventually supplanted that of the stop-cylinder. The improvements, said Napier, consisted 'first, in keeping two printing cylinders revolving in one direction by their being acted upon by reciprocating racks, and by each other alternately'. The second revolution occurred as the cylinder rose to permit the return of the bed carrying the forme.

Napier is credited with building the first printing machine to be established in Ireland, for the *Dublin Evening Post,* although it was manually operated. His cylinder machines were also the first to penetrate the North American continent. They were eventually imitated in the U.S.A. by Hoe, Taylor and others, and until 1850 were almost the only kind used in that country. In Joel Munsell's statistics of printing offices in Albany (1850), among the printing machines (as opposed to hand presses) in nineteen offices *Napiers* were in the majority, but how many of these were London-built it is impossible to say.

Napier's 'grippers' were a great step forward as they did away with the clumsy tapes or strings for conveying a sheet around the cylinder. Napier was also the first to make machines with small-size impression cylinders, and to devise toggles for bringing the cylinders down to print on the forme, and for raising them to let the forme run back without touching.

Between 1836–63 Napier made at least eighty-seven perfecting machines at prices between £450 and £600, to be driven either by steam or by hand. Ingram and Cook, publishers of the famous *Illustrated London News*, bought one in 1844. Napier also made newspaper machines from 1822 onwards, called the *Desideratum* (single cylinder) and the *Double Imperial* (two-cylinder). These were not as refined as the *Nay-Peer*. In 1838 he built a quadruple machine for the New York *Courier Enquirer*, but the year before had designed a rotary press (see Chapter 13), for some reason taking the matter no further. Under Napier's son, James Murdoch Napier (1823–95), the firm, in fact, replaced its cylinder-machine business with that of bed and platen machines for a different class of customer, as outlined in Chapter 8.

As the engineering industry developed, the manufacture of printing machines was undertaken by a variety of firms in Britain, the U.S.A. and on the Continent. Increasing efforts were made to improve performance, much attention being given to the smoother and faster movement of the bed, until the maximum speed permitted by the reciprocating motion was reached. The improvement in the feeding and delivery of sheets, in the inking and register and in the elimination of manual operations, where possible, were constantly sought. It is not possible to give details of every product of every manufacturer who arose in this new productive period, but an effort will be made to outline the major developments in each of the major manufacturing countries.

In Britain, the period from 1830 to 1860 was one in which many patents were taken out, but in which few machines made any permanent impact. Fairly successful was Robert Gunn, of Edinburgh, whose machines, in fact, were made by the engineers, Claud Girdwood & Co., of Glasgow. Gunn was responsible for installing the first steam-driven machine in Ireland for P. D. Hardy, a Dublin printer, in 1833, and followed with a double machine for the *Dublin Evening Mail*, which was claimed to print at about 2,300 impressions an hour. A Gunn machine printed in 1835 *Chambers's Edinburgh Journal*, the *Historical Newspaper* and *Information for the People*. It was designed to run at 750 impressions an hour and cost £300, without the steam-engine.

Most machines were intended for newspaper or bookwork and not for 'jobbing' printing. Napier tried to fill this gap with his *Desideratum* single-cylinder machine, although he had originally intended it for newspaper work. It was later taken over by Hopkinson and Cope. Two men could work it and dispense with steam-power. Movement was obtained from underneath by a mangle motion. The horizontal driving shaft carried a pinion, which geared into an intermediate spur wheel, which, in its turn, geared into a larger one at the end of the axis on which the impression cylinder was fitted.

Another engineer who successfully penetrated the jobbing field was Henry Ingle, whose family business had been in Shoe Lane, London, since 1820. H. Ingle & Co. pioneered what was originally a simple, light, single-cylinder jobbing machine known as the *City*. It was claimed (and this seems to have been the case) that it took less power to drive than similar machines. The exact date of its development is not known but in publicity in 1894, when a thousand machines were said to be in use, it was calculated that the *City* machine followed the *Belper* of 1835 but was prior to the *Wharfedale* of 1858.

The *City* was provided with a fly-wheel and handle for hand operation and pulleys and striking gear for steam-power. Sizes ran from Demy (6 × 4 feet) to 'Full size News' (11 × 7 feet) in a price range of £65 to £180. The popularity of this machine was such that some of the larger London printers installed them in batches, Waterlow's Finsbury works having thirty-five in operation at one time. Frederick Wilson in his *Typographic Printing Machines and Machine Printing* (1879) gave the machine a back-handed compliment when he wrote: 'Messrs. Ingle's machine is, perhaps, only remarkable for the absence of any complicated movements. It is of very light contruction, and is peculiarly adapted for printing light work, such as Government forms, &c. In consequence of its simplicity of make, it may be run at a greater speed than any of the single cylinders before mentioned.'

61. Ingle's improved machine

The comparison was with machines of the *Wharfedale* type, which followed the *City* from 1858 onwards. Ingle's did not accept the implied criticism. Their attitude was that the *City* had not originally been built to cater for heavy block-printing work. They went ahead to improve the inking, delivery and general structure of the machine to make it a good all-round press, capable of illustration printing, but retaining its simplicity of working. Despite its origin, therefore, as a jobbing press, the *City*—mostly in its 'Full size news' model (which by 1894 had gone up in price to £225)—was widely used by local newspapers.

Thomas Main, a printer on the *Morning Chronicle*, also noted the lack of a jobbing cylinder, and in 1850 took out a patent for his *Main* or *Tumbler* machine. The peculiarity of the *Main* consisted of the rocking action of the cylinder which, instead of revolving completely and remaining stationary, 'tumbled' back as the forme returned to be ready for the next impression. At the same time it was raised slightly from the bed to allow the forme to return. The advantages over the earlier kind of stop-cylinder derived from the speeding up of delivery. The *Main* machine was significant as the forerunner of the well-known *Wharfedale* class of printing machine, by which it was later superseded.

In 1854 Main joined with William Conisbee, who had his printing-machine manufactory in Southwark, but Main is said to have got the business into financial difficulties and to have fled to Australia. This does not seem to have held Conisbee back as he bought out the patent, improved the machine, and in ten years sold some 700 models. The speed was from 1,200 to 1,300 impressions an hour.

The firm of Harrild & Son (noticed as the pioneer roller manufacturers and which developed into a major printing supply house) made large purchases of *Main* machines from Conisbee, and in 1860 were claiming that nearly 400 were at work in different parts of Britain. Prices ranged from £70, for a fast jobbing machine, to £220 for one of newspaper size. The machine had thus developed from Main's original conception. This was not only a question of size, as Harrild's also advertised, besides the single cylinder, a 'new patent two colour printing machine' based on Main's patent.

At this point, because of the involved situation with regard to various machines, it would be as well to introduce a remarkable figure, Samuel Bremner, a North country journeyman who became printing manager of Petter and Galpin, whose works were at La Belle Sauvage, in Ludgate Hill, London. While in this position he projected the *Belle-Sauvage* machine, based on the *Main*, but differing from it as far as the motion of the impression cylinders was intermittent, remaining stationary when the sheet was being removed and while the succeeding one was being laid on. The *Belle-Sauvage* was lighter than the *Main*, and was said to be preferred by country printers, although it was slower in operation.

Conisbee was first called in to construct the *Belle-Sauvage*, but eventually Petter and Galpin felt that their real line of business was printing and not machine manufacture, and asked Harrild's to take over the machine. Bremner therefore transferred his services to Harrild's in 1863, and contributed considerably to that firm's success as they launched into the manufacture as well as the distribution of printing equipment.

Bremner was involved with a number of machines, and was a central figure in the development of a particularly British machine, known by the general name of *Wharfedale*. Basically, it consisted of an impression cylinder mounted on parallel side frames, a bed which, with the ink slab, moved to and fro, carrying the cylinder in gear one revolution, when travelling outwards, and leaving the cylinder stationary on returning, in order to admit the sheet being laid into the grippers for the next impression.

William Dawson was a joiner and cabinet maker of Otley, Yorkshire, in the valley of the river Wharfe (hence, eventually, the *Wharfedale*), who had been encouraged to make printers' equipment. In 1854 he was introduced to Stephen Soulby, a printer of Ulverston, by J. M. Powell, founder of the *Printers' Register*, and to the ubiquitous Samuel Bremner. The introduction arose as in 1852 Soulby had patented a printing machine called the *Ulverstonian* and two separate ironworks having failed to make the machine for him he turned to Dawson. The *Ulverstonian* was of the kind where the type forme is fixed to a stationary bed while cylinder and inking apparatus, in a travelling frame, rolls to and fro over it, pressure being obtained by the weight of the cylinder. This was not a very advanced idea as Norris had used it for the *Belper* in 1835.

Nevertheless, the *Ulverstonian* was made and marketed, and the connection between Soulby and Dawson lasted until 1859, and would probably have continued longer had not David Payne, Dawson's foreman, expressed disapproval of the principle on which the machine was built, and demonstrated the fact that the way to ensure a firm, sharp impression was to have the cylinder placed in fixed bearings and the type bed movable.

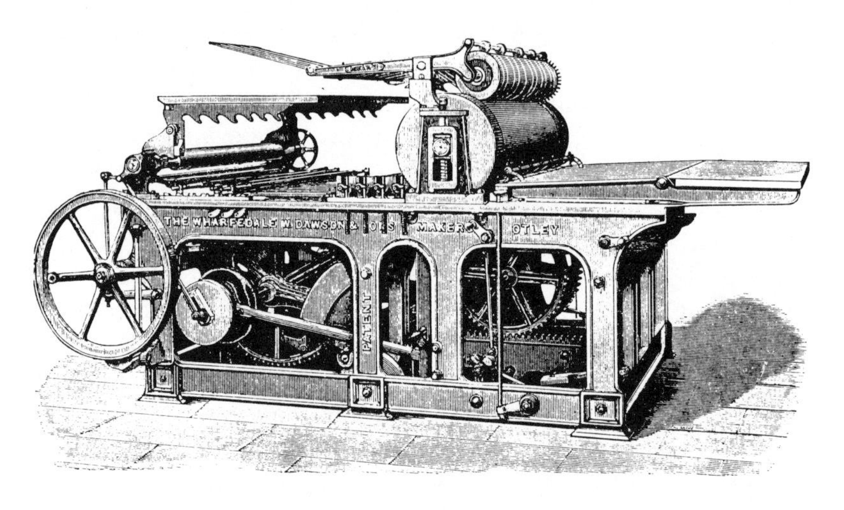

62. An early Dawson Wharfedale machine

Dawson and Payne therefore decided to build a machine on this basis in the Ashfield foundry which had been erected in 1854. The machine, which developed into the *Wharfedale*, was first called *Our Own Kind*, a name which adhered until 1860 (Plate XXXVII). The first *Our Own Kind* machine was delivered to G. W. McLaren, of Glasgow, and the first newspaper version to a Mr. Milner, of Lancaster. The machine was successful, and Payne was taken into partnership, the new firm being known as Dawson, Payne & Co. This partnership was dissolved in 1866, and the firm continued to be run by a succession of Dawsons. In 1874 various devices were introduced into the two-feeder *Wharfedale* to print different-sized sheets without altering the surface of the cylinder, and a perfecting machine, with a patented flyer, as well as a platen machine, were put on the market. A later machine, the four-feeder *Wharfedale*, came too late in the day, and was superseded by web-fed machines.

Payne, the actual inventor of the *Wharfedale*, founded Payne & Sons, and set up his own works. He secured a patent in 1867 for an idea which was incorporated in his version of the machine, which was known as the 'push back motion'. The stop of the brake wheel was brought into open space, thereby avoiding the jar and bump, and requiring less brake power than before. The same motion allowed the gripper points to pass the edge of the sheet before opening, thus enabling the feeder to place one sheet close after another without having to wait until the grippers passed. The cylinder was then pushed back into position for taking the next sheet, and nipped there, thus securing perfect register. When Payne's patent expired the ideas were applied to rival machines.

As Bauer, Hoe and others were to discover, skilled men tended to leave, and rival firms were founded by those who had learned something about the manufacture of the product. The two firms of Dawson and of Payne were added to in 1865 by another started by the brothers Elliot, who had held the contract to make the castings of the original machine, and by a third breakaway firm, Elliott & Co., in 1887. Various names were given to the presses manufactured, such as *Reliance*, *Caradoc* and *Defiance*, but they were all based on the so-called 'Otley principle' of the stop cylinder.

In 1863 Richard Watkinson joined with several men who had learned their trade at Dawson's and with Samuel Bremner, Harrild's manager, to found the Bremner Machine Co. at Otley, securing the backing of Harrild's. This became yet another manufacturer of the *Wharfedale* type of machine, which was called the *Bremner*.

Bremner had also seen the *Dutartre* machine, devised and patented in France in 1852—and so, incidentally, had the American, Hoe. Bremner admitted the superior inking power of the French machine, and designed the *Franco-Bremner*, adding a number of improvements. Eight inkers with riders entirely covered the forme, and the ink being supplied and distributed at two different points, excellent inking was obtained.

Hoe went so far as to attribute the invention of the stop cylinder to Dutartre when he introduced his version of the machine to America in 1853, after patenting a number of improvements. The Hoe machine printed at the rate of 3,000 to 4,000 impressions an hour and was nicknamed the 'little Astonisher'. The reason was that it surprised printers who saw it could do fine work, which they had thought only the hand press was capable of performing.

Indeed the attitude towards the cylinder machine was now changing. It was realized that it could, if well made, with better inking power, be used to print the better-class illustrated material. The *Wharfedale* seemed to be a good model to work from, but Joseph Parsons, printing manager of the *Graphic*, felt it had a weakness, particularly at the point of impression. To overcome this he conceived firstly the idea of providing a sound iron bed-plate, upon which the whole of the frame could be bolted, and secondly the girder under the cylinder, together with the runners, was strengthened—the whole ensuring an unyielding and firm impression. He co-operated with B. W. Davis, a former employee of Middleton, who had established his own manufactory in 1855, to build the *Graphic* machine which was to print the illustrated magazine of that name.

If Bremner could be inspired by France, France could be inspired by Parsons and Davis. A leading French manufacturer, Pierre Alauzet, who had set up in 1846, impressed by the *Graphic* machine, made a similar one for printing *L'Illustration* in Paris.

The *Wharfedale* continued as a model. The *Quadrant* machine, based on the *Wharfedale*, was originally introduced as a small jobber, but such was the demand that Powell & Son, the makers, began to offer the press in larger sizes. The bigger machines were fitted with a double-inking motion, and a taking-off apparatus, consisting of a cylindrical flyer which delivered the sheet printed side up without the use of tapes.

Newsum's, of Leeds, brought out the *Anglo-American* machine, based on Hoe's version of the *Dutartre*, but more massive in construction. At least six inkers were made to cover a full-sized forme, and arrangements were included for the accommodation of a series of riders. Furnivall's manufactured at Reddish an 'improved' *Wharfedale*; the Birmingham Machinists Company a version called the *Leader*; George Mann, of Leeds, the *Climax*; Pullan, Tuke & Co., of the same city, the *Cambrian*; and Alex. Seggie & Son, of Edinburgh, the *New Edinburgh*. The *Leader* dispensed with the large wheel and connecting rod, and the tables were driven by a horizontal driving disc placed immediately underneath. This was not a new idea, but had been abandoned earlier owing to the 'play' which occurred after very little wear. The Birmingham manufacturers therefore patented a guide, fitted with double-action anti-friction runners, by which any side movement was prevented.

While the *Wharfedale* was not the first to use a reciprocating bed—as Koenig, Cowper, Napier and Hoe had all acted in this direction—the Otley machine set the pattern for printing machines, particularly those used in Britain and its dependencies for some time. H. L. Bullen, an Australian, who became an American, writing in *The Inland Printer*, in February 1922, recalled that the first cylinder press he had worked on was an early *Wharfedale*. 'Three stout lads operated the press, one to turn the power wheel (as ours was not a "steam printing house", and gas engines and electric motors had not appeared), another lad to feed and a third to "fly" the sheets. We took twenty minute spells at each station. Now the *Wharfedales* have automatic flys, and printer boys know too much to substitute for motive power.'

Eventually, after the 1914–18 war, three of the firms in Otley amalgamated under the name of Dawson, Payne & Elliott Ltd., which began to diversify its output, as already by the 1890s the Otley engineers were losing ground to American competitors who based their machines on the two-revolution principle.

Owing to the rate of social and industrial advance in Britain it was in that country that much of the development in printing-machine manufacture took place in the first half of the nineteenth century. But the cylinder printing machine had been the invention of two Germans, Koenig and Bauer, who had taken advantage of the mechanical resources of London for their first machines, but who naturally took their plans with them when they returned to their native land.

As already mentioned, Koenig had returned to Germany in 1817, to be followed the next year by Bauer accompanied by an English fitter. Bauer had stayed in London to supervise the last improvements to *The Times* machines, and had brought with him an order for modifying them yet again, but owing to the difficulties at Oberzell, the job did not make progress. Despite these difficulties, the partners built the first German printing machine, a perfector for Johann Spener, of Berlin, in 1822, to be followed by three others. On 25 January 1823, the

Spenersche Zeitung became the first newspaper to be printed on a cylinder machine in Germany. Further orders followed from Cotta, Matzler and Brockhaus.

Freiherr Johann von Cotta, publisher of the *Allgemeine Zeitung*, Augsburg, had to be encouraged by the Crown Prince Ludwig of Bavaria to order a perfector, but once he did so he gave the press the name 'Schnellpresse' (rapid printing press), a name which has adhered to cylinder machines in Germany ever since.

France proved to be a good market and Guyot and Scribe, of Paris, bought a trial machine sent there in 1828. Between 1828 and 1830 no less than twenty machines were sold in France; and others to various countries, including Russia. Many of the Koenig and Bauer machines were broken up by angry pressmen during the troubles of 1830.

After the death of Koenig in 1833, Bauer was the master of the Oberzell works, but his leading collaborators began to leave and found their own firms. Koenig's nephew, Fritz Helbig, left in the year of his uncle's death and joined with Leo Müller, an Oberzell pattern maker, to establish the printing-machine works of Müller and Helbig in Vienna. A second Koenig nephew, Karl Reichenbach, who had installed the *Cotta* press, joined with his brother-in-law, Karl Buz, in Augsburg to found the predecessor of the Maschinenfabrik Augsburg-Nürnberg AG.

The machines built by Müller and Helbig were equipped with a feed mechanism invented by Napier, and they replaced the double rake drive by crank motion, and for stopping the cylinder made a mechanism equipped with double 'eccenters', which reduced manufacturing costs. This was a challenge to Bauer and, in 1840, he produced a rotary motion action for driving the carriage. This type of movement, popularly known as the 'sun-dial', is a combination of three geared wheels. The wheel's outward-facing teeth engage in the inner gear ring of twice the diameter. The crank pin, which moves in the centre line of the large gear wheel, is linked to the carriage by a connecting rod. The object was to enable the forme of type to move backward and forward without shock, and almost without noise.

Further competition came from new firms founded in Johannisberg in 1846, and by Georg Sigl in Vienna. In 1861 another Oberzell foreman, Andreas Albert, started a factory for making printing machines in Frankenthal. A direct line of printing-machine manufacturers had therefore been founded unconsciously by Koenig when he decided to leave Germany for London. The firm generally and popularly known as Heidelberg, after the town in which it is situated, and which plays so important a part in printing-machinery manufacture today, was founded in 1850 by Andreas Hamm.

The first power-driven presses in New York, said to be those of Fanshaw in 1826, were of the Treadwell type, and so were not based on the cylinder principle. But a year earlier the proprietors of two newspapers, the *New York Daily Advertiser* (morning) and the *New York American* (evening), imported a Napier cylinder machine from England. One of the proprietors went to London to buy the press, to learn how it worked and to transport it to New York, where it was shared by the two newspapers. This machine gave Hoe the idea of making cylinder presses himself, and his first was of the continuously revolving drum type, with cam-operated stops for positioning the sheets and with coiled springs for cushioning the bed at each end of its stroke.

This first cylinder installation is, however, not mentioned in Hoe's *Short History of the Printing Press* (New York, 1902), which states that as news of English inventions were reaching New York in 1832 Robert Hoe sent a young man, Sereno Newton (afterwards his partner), to England to investigate and to see what improvements were worthy of adoption. The first result of Newton's reports was the construction of three machines—the 'Single small cylinder', the 'Double small cylinder' and the large cylinder 'Perfecting press', it being admitted that these were primarily Napier inventions. As is the case with other presses, a name adhered to the type of press, and for a number of years in the United States all cylinder machines were known as 'Napiers'. Hoe added improvements from time to time—such as patented sheet fliers and, as early as 1847, a new bed-driving mechanism, known as the 'double rack and sliding pinion' bed drive. The reversal of the bed was accomplished by a roller at either end, entering a recess in a disc on the driving shaft, which in a part revolution brought the bed to a stop, and started it in the opposite direction.

It has been noted that shock absorption and limitation of noise was greatly desired, and a pioneer step in this direction was taken by Alva B. Taylor, of New York, formerly a foreman at Hoe's, when he patented in 1846 an air plunger and cylinder to cushion the bed at the end of its stroke.

The first printing machines used in France were of foreign manufacture, beginning in 1823 with a two-cylinder made by Applegath for the *Bulletin des lois*, Paris, and a 'presse à gros cylindres' for Firmin Didot. Two years later the *Journal des Débats* and *Le Globe* were printed on Napier drum cylinder machines. A general printer, Fournier, of rue de Seine, ordered a large Applegath cylinder in 1827, and by that year there were twelve printing machines at work in Paris, eight of which were of English origin.

The group of manufacturers which had welcomed the *Stanhope* and *Columbian* hand presses and had copied them, now turned their attention to the cylinder machines. The first practical French machine was that constructed by Gaveaux in 1831 for the journal *Le National*. This was a two-cylinder machine. In 1834 the mechanic Rousselet, of Paris, joined in the race, and by 1840 construction of printing machines was under way in Paris. Among the important names are Alauzet, Briard, Coisne, Derriey, Durand, Voirin, Dutartre, Normand, Thonnelier, Parrain, Gaigneur and Wibard.

Auguste Hippolyte Marinoni, a former apprentice of Gaveaux, aided by his former master, constructed in 1847 the first French four-feeder cylinder machine for the journal *La Presse*. He set up on his own in 1854, and was to become one of the world's most important printing-machine manufacturers.

The *Journal des Débats* eventually turned to a native engineer, and that intriguing figure, M. Selligue, according to Frey, produced a two-cylinder machine for the newspaper.

The Belgians were a little late in the development of their own printing-machine industry, the first factory being established by H. Jullien in 1862.

Russia had only one printing-machinery manufacturer before 1917—the firm of I. Goldberg, which, from 1882, began to make medium-sheet-size cylinder machines, in thirteen years producing about 300. After the Revolution there was no development until 1931 when the Rybinsk engineering works produced its first stop-cylinder machine, the *Pioneer*. By the outbreak of war (1941), 572 had been manufactured. The plant also made a large-sheet two-revolution machine—180 by 1941.

The method of feeding sheets of paper to a cylinder press differed according to the machine in question, and up to a certain point, according to national preferences. Drum cylinders were generally fed from the top of the cylinder and the printed sheets delivered to a table at the same end as the feed table. On stop cylinders the sheet was usually fed under the cylinder but delivery could be at either end of the machine. In the early perfecting machines sheets were delivered to a taking-off board under the receiving drums and between the printing cylinders. In some machines, such as Newsum's, the taking-off boards were just below the laying-on boards.

In America it seems that for many years feeding to the top of the cylinder was preferred, so much so that the Campbell *Country* press, which was fed under the cylinder, was considered unusual in the 1880s.

Front-delivery of sheets—that is to say in such a way that the sheet is brought forward in the direction of the cylinder motion with the impression side uppermost—was easier of achievement on the two-revolution press than on the stop cylinder.

Various types of drive were in use for moving the bed. It could be on wheels running in grooves, or be moved backward and forward by a crank, lever and geared segment working in teeth in its base, or perhaps, the commonest for a time, by a geared crank movement. The sun-dial of Bauer had its followers, particularly in France.

The inking apparatus might be such that it provided a downward flow from fountain to forme or it could be a lateral flow. Basically, a fountain is a trough or receptacle, which discharges ink on to distribution rollers which keep the ink in perpetual motion and on to forme rollers which actually apply the ink to the forme.

Each of these aspects of the press was capable of improvement and, as time went on and the cylinder machine became more effective and more widely used, printers sought increasingly efficient machines, designed to print in more than one colour and on both sides of the paper, and finally looked to the manufacturers to produce ancillary equipment to cut down the tedium of feeding and taking off sheets of paper. In a word, they looked for a completely automatic machine. It took some time for this to be achieved.

THE "MINERVA," Better known as THE "CROPPER,"

IS THE ORIGINAL PLATEN MACHINE, OF WHICH OTHERS ARE BUT IMITATIONS.

More than 9,000 of these Machines are in Use.

For SIMPLICITY, PERFECT DISTRIBUTION, REGISTER, EXCELLENCE of WORKMANSHIP, and STRENGTH of all the Parts the "Minerva" is without a rival.

The IMPRESSION may be REGULATED by a SINGLE SCREW, or thrown off and on instantaneously.

Messrs. McCorquodale & Co. have 14 Minervas.

The following are a few EXTRACTS from TESTIMONIALS:—

"I have a lad printing two thousand an hour on one of them."—Chas Eagle.

"No jobbing office is complete without your economical and unique machine."—Curtis Bros. and Towner.

"I would not exchange your machine for any other which is in the field."—Samuel Johnson.

PRICES AND TERMS ON APPLICATION TO

H. S. Cropper & Co., Great Alfred Street, Nottingham.

London Depot: 33, ALDERSGATE STREET, E.C.

10

THE JOBBING PLATEN

POSSIBLY THE MOST popular printing press during the last hundred years has been the ubiquitous 'jobbing platen', which can be seen in various forms at both ends of the industrial scale—in the biggest general printshops as well as in the garages and even dining-rooms of amateurs. There are operatives whose damaged fingers testify to hasty work on an unguarded jobbing platen, but who still recall with nostalgia the first *Cropper* on which they worked, using a generic term in the same way as an earlier generation referred to the *Eagle* press.

'Job' printing was originally considered to be any work which made less than a sheet, although today the word generally encompasses anything not considered a book, periodical, newspaper or specialist production such as a package. For the printing of small items the normal-sized common or wooden press was too large, unless a number of jobs could be accumulated and printed up together—an inconvenient proceeding. Printers' joiners constructed small-scale models for individuals, but these must have been expensive and, in any case, there was always the drawback of hand-inking. In this respect the iron versions of the hand press were no improvement, and the few self-inking contrivances which were devised needed the application of extra manual power on the part of the pressman. Small models of the iron hand press were made, some being called specifically 'card' presses, and were used by both professionals and amateurs. But something faster and cheaper was required by the printer at a time when opportunities for job work were increasing.

Before about 1830 letterpress job work was limited. Bills were handwritten and personal and trade cards were mostly engraved. Then trade and industry began to expand and there was a consequent demand for billheads, business cards, handbills and advertising material, as well as for personal stationery—a demand stimulated by the invention of the envelope and the extension of postal services.

The answer to the small printer's requirements was the jobbing platen press, often now called simply a 'platen'—like the 'bed and platen' it derives its name from the flat pressing surface of the original wooden press.

The jobbing platen was very much an American development and the earliest known experiment, which indicates the evolution from the hand press, was made early in the nineteenth century by Daniel Treadwell, of Boston, Massachusetts, who has been noted as the inventor of the first power-driven platen machine. In 1818 he built, but did not patent, a press worked by a foot treadle instead of the customary hand bar. This press was found to occupy too great a space—that of three ordinary hand presses—and was not successful. Thereupon Treadwell decided to go to England, being possibly encouraged by Clymer's example, and left for London in November 1819. The next year he took out a British patent for 'certain improvements in the construction of printing presses', which constitutes the forerunner of the jobbing platen. Treadwell's aim was to utilize the pressman's leg and weight of his body to take an impression, and this was to be achieved by using a treadle. He also sought to eliminate the need to run the forme under the platen and to incorporate a device by which the pressman could turn a sheet if he wished to print on both sides from the same forme.

Treadwell arranged for his press to be manufactured by Baisler and Napier, Lloyds Court, Crown Street, Soho, but, according to Hansard, only one press of this construction was made in England. As there seemed to be no support for his treadle press the inventor returned to Boston, where he devoted himself to the bed and platen machine. There is no known surviving example of Treadwell's treadle press, but some idea of its working can be gained from the patent. One aspect may be noticed in passing—the use of a great lever on a pivot, reminiscent of Clymer's *Columbian* press, with which Treadwell could have been acquainted.

Treadwell's press consisted of a cast-iron frame with two sides united with cross-pieces; and a large beam or lever (with a spring on its top), which moved on an axis near the head of the press. From the shorter end of the beam the platen was hung from a circular cap into which fitted the tops of a number of bars set in a pyramidal shape—a rigid rod in the centre and the others with their lower ends resting in concavities. Under the platen was a table carrying the forme. From the end of the longer projecting portion of the beam a hinged bar connected with a treadle which projected in front of the press. Depression of the treadle caused the bar to straighten up and push the longer end of the beam upwards and, hence, the shorter end and the platen downwards to make the impression. When the treadle was released the weight of the larger part of the beam, aided by the spring, brought the parts of the press back to their former positions, at the same time releasing the top of the platen bars from their cap to allow the platen to be thrown up in order to feed with paper and to ink the forme. The platen was counterbalanced by a weight and springs and turned on an axis, resting in the frame, and could be held by a rod hooked in the lower frame. There were two common friskets which could revolve, the idea being that the pressman could turn the sheet on the second side in printing half-sheet fashion. As to the performance of the press, Hansard writes: 'There is certainly great originality in the construction of this press: its operations are conducted with much facility by one man; and as the rolling of the table, and the horizontal movement of the bar, are dispensed with, the labour must be considerably reduced.'

While the idea of a hinged platen and the treadle operation provided elements for the later jobbing press, it will be noticed that Treadwell proposed nothing novel in the way of inking, and that he continued to use the traditional frisket for holding the paper. Hansard's footnote: 'Query. Did the name induce the invention, or the invention the name?' is soon answered. The word 'treadle' for a foot-operated lever dates back to the fourteenth century.

However, back in the United States Treadwell did contribute an idea which was eventually taken up by the manufacturers of jobbing platens, although he devised

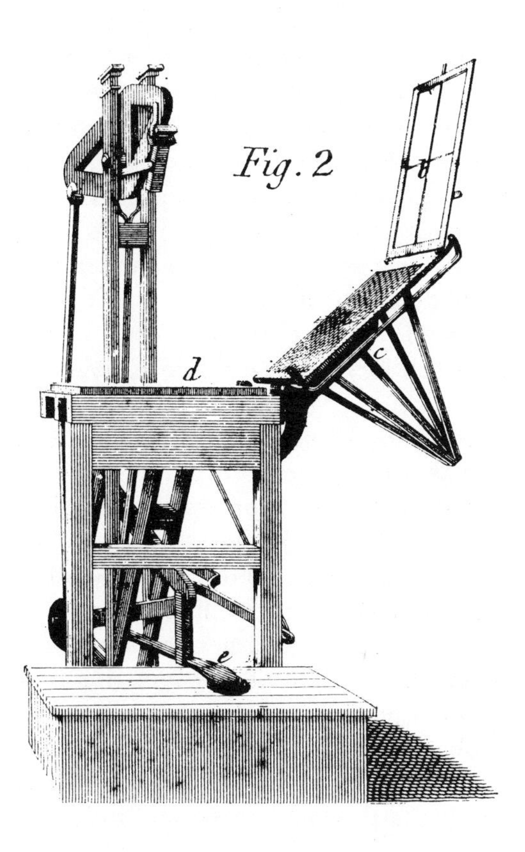

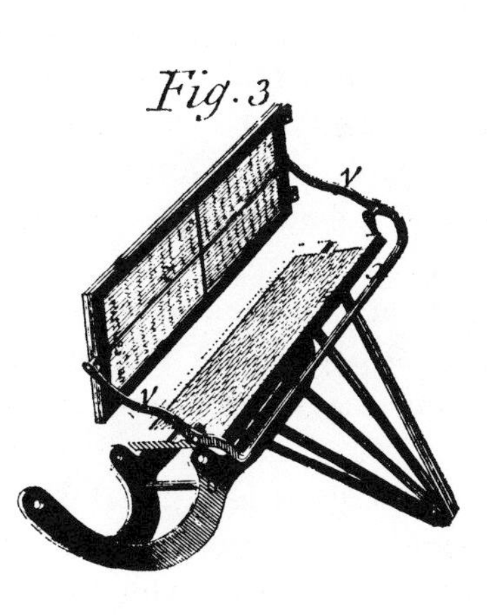

64. Treadwell's press (See page 144). Fig. 1—the platen in printing position; Fig. 2—the platen released and thrown back; and Fig. 3—a revolving frisket

it for an entirely different kind of machine—the 'bed and platen' press, the first of which was built in 1821. The novelty was the rotating ink disc. Treadwell did not originate this mechanism, because Sir William Congreve had developed it between 1819 and 1820 for use on his security printing press in London.

In Britain at least one inventor conceived the idea of providing a vertical forme with a platen or printing surface brought into contact with it by means of vibrating arms moving on pivots. The inventor was John Kitchen, a printer of Newcastle upon Tyne, and his patent is dated 1834. He proposed obtaining the power to apply the pressure by means of jointed levers actuated by a crank and rod. Nothing is known of this machine, and it is unlikely that it would ever have become popular. Compared with the eventual jobbing platen it was a complicated piece of mechanism. If a model was made it must have been an interesting sight as Kitchen seems to have been influenced by the Gothic revival. His press was to be contained in a frame reminiscent of a piece of church furnishing.

The characteristics of the jobbing platen press as it emerged are that the type forme and platen are pressed together by a horizontal rather than a vertical movement; that no movable carriage is needed and that inking is automatic. While the original intention was to release the pressman's hands by use of a foot-operated treadle, in later table-models the pressure was applied by a hand lever. The treadle, in any case, ceased to be a special characteristic after the application of power—originally from steam-engines, then by internal-cumbustion engines and eventually electric motors.

The jobbing platen, therefore, developed in two directions—towards a small, hand-operated press for the amateur, and towards a power-driven machine with which professional printers could ultimately challenge the cylinder press. As will be noted, different approaches were made, over the years, by manufacturers to bring the bed and forme and the platen together. At first both bed and platen were movable, as in the Gordon and derivatives; whereas in 'clamshell' models the bed is fixed in an upright position against which the platen rotates. In later presses, such as the Heidelberg, a knee-lever motion brings the platen against the bed and forme.

Certain old terminology was retained, the platen being a steel slab covered with sheets of paper, over which a final sheet is stretched; the tympan to describe the packing and its enclosing frame, and the frisket for the fingers which hold the paper.

Treadwell was followed by a fellow Bostonian, Stephen Ruggles, who after making a cylinder press in 1827 tried his hand at a card press in 1830. Neither press was manufactured, owing to lack of capital, but in 1839 he completed his so-called *Engine* press. This was the first self-inking, treadle-driven job printing press, but the bed and platen were still horizontal—with the type forme above and the platen below. The traditional frisket was eliminated and the ink was distributed by a roller. A speed of 1,200 impressions an hour was claimed, and a new era seemed to be opening up for the small printer. However, the position of the type was precarious—it could become loose and fall on to the platen, and the mechanism was too complicated, so much so that Ruggles probably felt that a simpler construction was desirable. He therefore took his idea a step forward and constructed a new press, known as the 'card and billhead' press in 1851, which can be said to be the first press with a vertical bed. It had a printing area of only $4\frac{1}{2} \times 7\frac{1}{2}$ inches. While the bed was vertical it consisted of the flattened side of a fixed cast-iron cylinder, the axis of which was horizontal. The remainder of the cylinder formed an ink-distributing surface around which ink rollers travelled. The platen oscillated to and from the bed by means of a connecting rod attached to the cylinder.

Ruggles produced a number of jobbing platens, including the *Diamond* card press, with a chase of only $3 \times 4\frac{1}{2}$ inches, which could stand on a table. This had ceased production by about 1860 but was the genesis of the idea that small platens of this sort could be manufactured for the use of amateurs. Curiously, some time later, the success of one such amateur press, the *Model*, encouraged its manufacturers to turn in the other direction and to enter the commercial field with an enlarged version.

While immediately following job platens failed to adopt Ruggles's idea of the flattened cylinder, it was not forgotten and it was made use of in presses made, for example, by A. B. Taylor of New York, and in England by the Birmingham Machinist Company and Furnival & Company, of Reddish, and is used in the contemporary *Adana* T/P48.

Ruggles-style presses gave way to the more popular productions of George Phineas Gordon, orginally a New York printer. No example of the card press he patented in 1850 is known to exist, but there is still an *Alligator* platen press of 1851, which is owned by the Nebraska State Historical Society (Plate XXXIX). The *Alligator* was so named as it was a dangerous press, with a fixed platen and hinged bed which tilted forward so suddenly that it was a danger to the feeder's hand. H. A. Brainerd, the printer who, in 1911, presented the specimen to the Society, stated that at least fifty men had had their fingers smashed while feeding it. The *Alligator* was developed before Gordon saw the value of an inking disc and the inking therefore was from a somewhat primitive curved and sloping metal plate. This style of inking plate was used later, in a more refined form, on the British Bremner jobber. Few *Alligators* were built and later in 1851 Gordon was granted the patent which was the basis of all his later platen jobbers.

From Gordon onwards the jobbing platen took on its familiar form with a vertical or slightly sloping bed and inking from either rotating discs or from rollers automatically connected with the motions of the platen. But there were and are variations in the application of the basic principles. Gordon's system was that of the bed and platen rotating towards each other in the manner of a closing hinge—the bed meets the platen, which moves into a near-vertical position for the impression and then retracts to a near-horizontal position for the feeding of the next sheet of paper. At first, his inking system was similar to that of the *Alligator*, and it was not until an improved version, produced in 1856, that Gordon rediscovered Treadwell's idea of a revolving disc to distribute the ink for the rollers, and he is credited with the invention of an improvement. This combined with a rotating disc an 'annular' disc 'which shall revolve around and in a contrary direction', to use the words of his patent. This improved the inking as the rollers passed over an entirely new surface every time they reached the discs. Gordon was a spiritualist and maintained that the mechanism of the press had been described to him in a dream by Benjamin Franklin. He therefore named his improved model the *Franklin*. Despite this, the press was more generally known as the *Gordon*, particularly when its principles were followed by other manufacturers. Nevertheless, Gordon captured so much of the market that even the formidable Hoe & Co. gave way to him. In 1861 they took up Franklin L. Bailey's patent for a jobbing platen, which was called the *Caxton*. It would print a sheet 9 × 12 inches, but was never a favourite, owing to the dominance of the *Gordon*, and its manufacture was discontinued.

65. The Franklin or Gordon press

Gordon brought out a new-style press in 1872, in which the platen and bed were hinged together and which incorporated a 'throw-off'. The press was not as popular as the earlier model. The throw-off is a device which limits the motion of the platen when the operator misses a sheet. It is operated by a lever at the left-hand side of the press between platen and fly-wheel. This device was first mentioned in a patent granted to James Young, of Philadelphia, in 1852 and also by Ruggles in one of his patents, without claiming it as his own. It is possible that he made arrangements to use the Young invention. A differently conceived throw-off was later used on the *Liberty* press, the inventor, Degener, claiming it as his own. Gordon did not provide a throw-off on his early presses but by 1872 when he brought out his new-style press the Young patent had expired and he was therefore able to include this useful device.

A variation of the Gordon principle was used in the *Liberty* or *Degener* press, invented in 1859 by Frederick Degener, who had been Gordon's draughtsman. The bed and platen rotated towards each other on a joint axle. With the forward movement bed and platen closed and with the reverse movement the forme was inked. This press was very popular and, according to an official report, was the most-used treadle platen in Belgium. But the presses may not have come from America, for, in about 1881, Degener's former partner, F. M. Weiler, established a factory in Berlin, and after United States manufacture was discontinued in 1890, *Liberty* jobbers came from Germany and continued to be sold until at least 1914.

Another system incorporated a fixed bed against which the platen rotated. An example was the *Pearl*, a light, fast jobber, made by Golding & Co., of Boston, from 1869, and the *Golding* jobber, a more elaborate press, invented in 1880. The *Golding* art jobber had a more advanced inking system, which combined both cylinder and disc. The *Pearl* was ideal for short runs of small items; the art jobber was a more ambitious press for fine colour work.

An improved *Franklin* was shown in London at the International Exhibition of 1862, and in 1867 H. S. Cropper & Co., of Nottingham, commenced its manufacture under the name *Minerva*. But Cropper's name became so attached to this type of press that any jobbing platen press operator was known as a *Cropper* hand, the term occurring in the minutes of the Platen Machine Minders Society, which was formed in London in 1890, as jobbing platens had by that time become so numerous. As to performance of the *Croppers*, it was repeatedly claimed that a boy could work a treadle platen at the rate of 1,000 to 1,250 impressions an hour and even more, but this was probably an exaggeration if taken over a period of time. Only when the press was power-driven could this speed be achieved consistently.

Gordon's original patents ran out and variations of his press were made by Shniedewend and Lee of Chicago (the *Challenge*) and more especially by the firm of Chandler and Price, which, in 1901, bought the old Gordon works and the right to use the name '*Gordon*', although they had already produced *Gordon* old style platens. Their 'New series' platen was introduced in 1911 and then their *Craftsman* press for half-tone and fine work—both based, however, on the original Gordon principle. The Chandler and Price presses were sold in Britain by H. W. Caslon & Co.

Variations in the manner in which the platen and bed came into contact characterized different makes of jobbing platen, but the inking systems also differed. Important in this development was the Gally press and its successors, the *Victoria* and the *Phoenix*, which had a stationary bed and a platen which rolled to a vertical position, then glided forward so that immediately before the impression the platen was parallel to the bed and moved perpendicularly towards it. The first of this type of press was invented in 1869 by Merrit Gally, of New York, and was introduced under the name *Universal*. But Gally also rejected the disc and roller method of inking, substituting a full-width fountain and distributors to transfer the ink to

66. The Challenge press. (See page 149)

a large drum from which it was picked up by the forme-rollers. The *Universal* could be adapted for embossing, stamping and creasing. After 1887 it was manufactured as the *Colt's Armory* press, and was the origin of presses of this kind. The *Sun*, manufactured by Greenwood and Batley, of Leeds, for example, resembles the *Universal*, with an inking apparatus consisting of revolving cylinders and independent wavers. Dawson's, of Otley, made the *Mitre* on the same lines, and Harrild's *Fine Art Bremner* platen had distributing rollers on the lines of the cylinder printing machine. These presses, therefore, extended beyond the field of card, billhead and leaflet work and were used for more extensive printing, particularly from heavy engravings.

In its development the comparatively humble jobbing platen played an important part in the transformation of the printing trade. It speeded the production of the mass of small items required by industry and commerce, and was an important factor in the abandonment of the practice of wetting paper before printing. The jobbing platen became so indispensable a part of printers' equipment that presses were manufactured in great numbers. By 1894 no less than eleven firms were manufacturing *Gordon* presses, and it has been calculated that during the century 1840–1940 no less than 123 different kinds of treadle-driven jobbers were made in the United States alone, bearing names such as *Baltimore*, *Washington*, *Favorite*, *Peerless*, *Leader* and *Star*. Names were changed for the European market, the *Franklin* becoming the *Minerva* in Britain, and the *Universal* the *Phoenix* when marketed by Schelter and Giesecke, of Leipzig.

The manufacture of others was transferred from America to Europe, as in the case of the *Arab*, which, at times, was known, rather oddly, as the *Anglo-American Arab*. In France M. Berthier imported American platen presses and then manufactured his own under the name *La Minerve*. Pierron et Dehaître's *Le Progrès* resembled the American *Liberty* and M. Wibart (*La Sans-Pareille*) and M. Poirier (*La Merveille*) frankly turned to America for their models. An official report of the Belgian Ministry of Industry in 1911 revealed that many of the platens operated by lever were of American origin, but also mentions two home-built models—one with a movable bed, made by Jullien, and another with a fixed bed, made by Carabin-Schildknecht. Neither favoured the disc system of inking. The only manufacturer of presses in Russia before 1931 was I. Goldberg, who commenced operations in 1882. He produced a jobbing platen, but on what lines is not known, but significantly in the report of the Five-Year Plan for the Printing Industry (Moscow, 1929), platens are referred to as *Amerikanki*. In South America jobbing platens are generally known as 'prensas *Minervas*'.

67. 'Presse à pédale à marbre renversable' made by Établissements H. Jullien. (See this page)

The British also marketed home-built machines. The *Bremner*, made by Harrild's of London, had a locked platen to secure 'dwell' on impression and to prevent slur. This machine used an unusual curved plate for inking, rather than the disc. J. M. Powell & Son's *Empire*, on the other hand, was characterized by an extra-large disc-like inking surface, nearly circular but cut across the base. The *Invictus* platen, made by the Birmingham Machinist Co., reverted to the Ruggles 'flattened cylinder'. Nearly the whole of the surface of the cylinder acted as the inking table and the flat

68. The Invictus platen for foot or power, manufactured by the Birmingham Machinist Company. (See page 151)

part remaining as the bed on which the frame was secured. The ductor, supplying the ink, pressed against a travelling rubber apron which, in turn, provided ink to the largest of five inking rollers. By means of cams the other four rollers were made to transverse the forme between every impression, the centre roller, with its sideways motion, facilitating the distribution of ink. More elaborate still was Godfrey's patent gripper platen machine, combining platen and rotary motion, which was made by Furnival & Co. The forme was placed on the flattened side of a revolving cylinder, the platen occupying the same position as on other jobbing machines. The lay was automatic—the paper being placed on a stationary board and being taken and adjusted to its exact position. The *Adana* T/P 48, manufactured today by Adana (Printing Machines) Ltd., is a quarto machine adapted for treadle or power, which utilizes the flattened cylinder. This constitutes another example of a firm developing from the amateur into the professional printing market. While the firm's smallest presses are for amateur use, the power platen, in particular, is used by small jobbing printers.

In 1893 Fred Harrild, of Harrild & Sons, conceived the idea of combining two jobbing platens to economize on space and motive power. The press was launched in the June of that year. The heavy iron framework carried at each end a platen press arranged in the usual manner, but the rocking frame which normally carried only one type bed in this model provided a rigid type bed on both sides, back to back, giving an impression on each plate alternately, and thus producing a double result for the same amount of power expended. The inking system was duplicated, but two ink ducts and two discs were provided to allow different colours to be worked simultaneously. Two operators could work independently of each other.

At first in America, the land of their origin, the sizes of treadle platens were based on a fractional part of a 'medium' size hand press. Presses having a chase size from 7×11 inches to 9×13 inches were classed as '$\frac{1}{8}$ medium'. From 10×15 inches to 12×18 inches they were called 'quarto medium', and from 13×19 inches to 17×22 inches 'half medium'. A certain amount of confusion was caused when announcements followed the nomenclature but omitted the dimensions of a press. A more practical system developed, therefore, whereby the size of the chase was given. As an example, the *Model* was available in the following sizes: 3$\frac{1}{8}$×5$\frac{1}{8}$ inches; 5×7$\frac{1}{2}$ inches; 6×9 inches; 7×10 inches (hand operated); 6×9 inches; 7×11 inches; and 9×13 inches (treadle operated).

Traditions died hard, and the practice of using decorative ironwork continued, in the case of the *Prouty* press with a flower motif. The Golding company, wishing to identify their *Official* press (a development of the *Pearl*), encased the working parts in thin sheet metal, shaped like a tulip and decorated with strapwork. Nothing quite as elaborate as the *Columbian* hand press was attempted, with the exception perhaps of the *Columbian* so-called rotary, which had no connection with Clymer's *Columbian* press, and which was not strictly speaking 'rotary'. It was a jobbing platen, made by Curtis and Mitchell of Boston, from 1878, and the cast-iron pedestal incorporated a portrait plaque of *Columbia*.

Some idea of the impact of the jobbing platen on the printing trade may be obtained from the comments of John Southward, who wrote in his *Progress in Printing and the Graphic Arts during the Victorian Era* (1897): 'Printers were slow in appreciating their merits, but as soon as they were understood, it was seen that they were going to effect important changes in the business. Previously work of this kind was done at a press, on which, with a boy rolling and a man pulling, 250 commercial cards could be printed in an hour. A quick platen machine will now do 2,000. The public used to be charged for 500 of such cards about 7*s*. 6*d*., but 1,000 of them are now done for 4*s*. 6*d*. Handbills, which in the old press days were about 5*s*. per 1,000 are now charged 8*s*. 6*d*. for 10,000 and so on. This has been brought about by the use of the small platen, which can be worked by one boy. But the printing is not only done more rapidly and more cheaply, but ever so much better. A foolscap folio machine is enormously stronger than a foolscap folio press of any kind. The latter gives a spongy, yielding impression, owing to the blanket between the tympans; the other gives a hard, unyielding impression, clear and bright. The public would not now tolerate the business cards, for instance, that were accepted sixty years ago. There must not be the slightest impression on the back, and with a press blanket this is unavoidable, to some extent, at least; while it is wholly unnecessary when a machine with hard packing is used. The comparative inexpensiveness of platen-machine printing has had much to do with the extended practice of colour printing. One thousand copies of a circular to be printed in four colours on a press would occupy two men, or a man and a boy, about a day. The work could be done by a boy alone in half a day at the machine. Besides the better impression of the latter, there is much better register, which is of great importance in colour work.'

When the bigger printers realized the advantages of the jobbing platen they installed them, but abandoned the treadle in favour of steam-power, as a quotation from S. H. Cowell's *A Walk Through our Works* (1888) indicates: 'A machine of this kind is a very compact affair, and may be worked by a boy, who can set it in motion by a treadle, worked something like a pedal in an harmonium. In this office, however, all machines are driven by steam. The machine prints at a rate of from 1,200 to 1,500 per hour, if the attendant is expert in laying the sheet to the gauge. These little machines are made to do very beautiful work when such is required. Several varieties of Jobbing Platen Machines are used in this office. There is a long row of what are known as *Minervas* (from Nottingham), as well as *Arabs* (from Halifax) and *Universals* (from Leeds). . . .'

The change from steam-power to electric motors stimulated the search for other improvements, particularly in feeding and delivery arrangements. An automatic lay had been developed for Godfrey's patent gripper platen in the 1880s and by 1913 the Miller Company had provided an automatic feeder for the Chandler and Price presses. The way was open for a completely automatic jobbing platen. A number of firms, particularly in America and Germany, began production. The Miller company which had made the automatic feeder for Chandler and Price, built their own presses into which they incorporated the feeder, working towards the automatic press—the *Master-Speed* automatic, which was launched in 1925. Brandtjeb and Kluge were pioneers in the manufacture of suction devices for both feed and delivery and introduced the idea of feeding from a stack of paper standing on its edge. Their automatic press was marketed in 1928. The German Heidelberg Company (founded in 1850) began its first trials with an automatic platen in 1912, and produced a press with a revolving arm for feeding and delivery (attributed to a printer named Gilke), which represented a new step forward in design. In 1925 the machine became known as the *Heidelberg* automatic platen. Some motor-driven presses retain hand-feeding which may be necessary for certain operations, but, for example, the Thompson British Automatic Platen and similar machines have a completely automatic feed and delivery of paper, locking register device and geared inkers.

During the 1880s the idea of a reel-fed platen occurred in America to Wellington P. Kidder and, in England, to Greenwood and Batley. Kidder designed a press on the lines of a *Gordon*, but with a fixed platen and a swinging bed. It was fed by a reel of paper, the web being cut into sheets by shears at the delivery end; and could handle any stock from tissue to cardboard. The *Kidder* self-feeding job press, as it was called, could print at a rate of 2,000 to 5,000 impressions an hour. Kidder may have got his idea from an 1853 invention of George Phineas Gordon called the *Firefly*, a new kind of fast card press. The stock fed into it was not in the form of individual cards but cardboard strip in roll form. The machine was operated by a hand crank and after the impression was made the strip was cut into individual cards. A similar press was patented in 1857 by Franklin H. Bailey, who assigned his patent to R. Hoe & Co. The fly-wheel and ink cylinder rotated on the same shaft which was turned by a hand crank. In 1865 the machine was adapted to include a numbering mechanism and was used to print theatre and railway tickets.

In Greenwood and Batley's machine the reel of paper was slung into position on the top by a small crane. The machine was provided with adjustable knives which cut the reel into sheets after printing. In America this development continued, although it is necessary to make a distinction between web-fed platen presses, which are derivatives of the jobbing platen, distinguished by the vertical positioning of the bed and platen, and the web-fed 'bed and platen' which derives from the machines of the *Adams* or *Napier* variety.

Kidder's invention was important because it pointed the way to multiple press units, which could print on both sides of the web, and to in-line operations after printing. This group therefore tend to consist of specialist presses for tickets, tags and office forms. Following Kidder, the New Era Co. brought out its first web-fed platen press in 1900, which could print up to 6,500 impressions an hour. The New Era multi-process press was introduced in 1950 and the most up-to-date prints three colours on the web by use of three platen units up to speeds of 7,500 an hour.

Over the years, improvements have been made to the more conventional job platen press—in ink distribution, in safety devices—until it has reached its highest point by which some models are capable of printing at 5,000 impressions an hour. With the development of other presses and techniques, however, the jobbing platen is entering a decline, although it will continue to be used for smaller work, and is still a useful part of the printer's equipment. Treadle-operated platens are used in places where there is no power-supply, and are also the standby of amateurs and schools, despite the hard work involved with the heavier models in 'kicking the press', as the treadle operation is called.

One development probably not contemplated by the originators of the jobbing platen is that its introduction led to the virtual abandonment of wetting the paper before printing. It is true that dry printing had preceded the adoption of the jobbing platen, as in 1849 William C. Martin, of New York, successfully printed from dry paper on a hand press, but few were able to imitate him. It was the view of Theodore Low De Vinne, the famous New York printer, that the use of the jobbing platen finally broke down the prejudice against printing on dry paper. Interviewed on a visit to England by the *British and Colonial Printer* (5 July 1894), he said that the invention of Gordon's treadle press had begun the practice of dry printing, to be followed by cylinder presses. It became apparent that dry printing on smooth paper was best, and that it led to a saving in time.

Trials were made with dry printing on power presses, but by the 1860s only the most skilful pressmen would attempt it. Ten years later the practice was widely adopted in America, but even into the period of the rotary press wetting paper was thought to be necessary in Britain and Germany. But eventually dry printing prevailed and although today a private hand-press owner will still damp his paper the practice had died out in commercial printing.

69. Machine room of P. Dupont, Paris, 1867

11

THE CYLINDER MACHINE TAKES OVER

MUCH OF THE development of the cylinder printing machine in the second half of the nineteenth century took place in the United States, and the continued expansion of that country kept up the demand begun in the hand-press era, for inexpensive, light and easily operated machines to be set up in the new territories and in country towns away from the big industrial centres. The quality of the work a machine could produce was often of secondary consideration. Many tried their hand at inventing and manufacturing printing machines to follow Hoe and Taylor, and while there were the inevitable failures a number went on to found established firms. The names of Potter, Newbury, Henry, Campbell, Scott, Pasko, Babcock, Cottrell and Prouty are but a few.

An early entrepreneur was L. T. Guernsey, of Rutland, Vermont, who constructed a cheap machine in 1852 known as the *Guernsey* Oscillating Cylinder press. Among his competitors was Joel G. Northrup, of Syracuse, New York, one of those fertile minds who had endeavoured to mechanize the hand press and who had experimented in a variety of ways, including one press with 'rotating platens'. In the same year as Guernsey he introduced his *Country* press and claimed it would print at a speed of 1,000 impressions an hour. Whether this claim was justified or not there is no doubt that his press was popular among country printers. George Gordon, the jobbing platen pioneer, was impressed and tried his hand in the cylinder field, and after an early experimental failure, produced a cylinder machine in 1856 which utilized the rotating ink disc, so familiar on jobbing platens. Gordon used it again on his *Oscillating Cylinder* of 1864. After his 'improved' oscillating cylinder of two years later Gordon ceased to manufacture cylinders, as he did not make so much money out of them as the jobbing platen, for which he became famous.

The bigger manufacturers soon became aware of the market provided by the smaller newspaper and job printer, and numerous *Country* presses were made. Somebody was bound to revive the old idea, for simplicity's sake, of a machine with a stationary forme and a lateral-moving cylinder. In England this had been tried as early as 1835 by Edwin Norris with his *Belper* press, and, as has been recorded, Stephen Soulby had eventually been unsuccessful with his similar *Ulverstonian* in 1852. However, in the United States the *Davis* 'cylinder' press (1885), the *Allen* 'hand press' and the *Vaughn Ideal* hand cylinder press were used by small country printers, with no facilities for power, for a long time, the *Vaughn* press being patented as late as 1892.

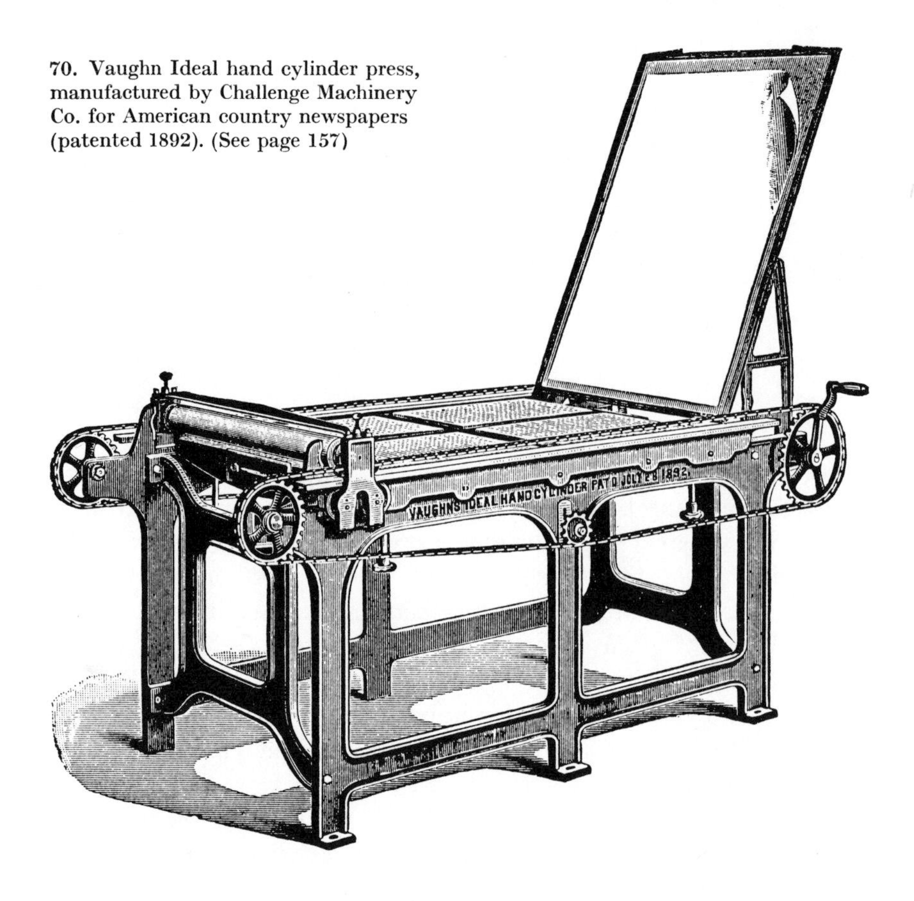

70. Vaughn Ideal hand cylinder press, manufactured by Challenge Machinery Co. for American country newspapers (patented 1892). (See page 157)

While presses were developed for specific kinds of work, such as the *Campbell Book* press, an early American two-revolution machine, there was the temptation in a period of increasing competition to claim that machines were capable of tackling many kinds of work. In 1886 Hoe offered their *Job and News* cylinder press—of the drum cylinder kind—which had a claimed maximum speed of 2,500 impressions an hour. The *Fairhaven* press, made by Golding of Boston, was for 'news, book and job work'. Publicity stated: 'This press is designed for printers whose offices have outgrown the capacity of the largest job presses, but whose work does not warrant the purchase of a high-priced cylinder, which are not only expensive in first cost, but require a first-class pressman to run them, while the cost of motive power, rollers, ink and incidentals is so great that they cannot profitably be run in the average job printing office. The *Fairhaven* press is simple in construction and therefore easy to understand and operate; is not liable to get out of order; can be run either by hand or steam power; requires but three rollers; and has an impressive throw-off, which greatly reduces the usual percentage of spoiled sheets. We do not claim that this press will turn out fine cut and registered color work in a style equal to a stop-cylinder press, with stacks of rollers, costing two or three thousand dollars; but we do claim that the *Fairhaven* will turn out first-class newspaper and book work, and will meet all reasonable demands to the satisfaction of the purchaser.'

The cylinder made two revolutions to each impression, rising slightly to allow the bed to pass under it on the return movement. It was claimed that the impression could be adjusted in a few seconds, as it was controlled by a single bolt at each end of the cylinder. The blanket and overlay sheets were held by separate shafts for convenience in making ready. The impression throw-off, which was operated by the pressman's foot, threw the cylinder out of gear to allow the bed to pass under it without giving it an impression.

But, naturally, in the major printing centres there was a desire to produce 'fine cut and registered color work' and there was a consequent need for a heavy-duty press. This is where the two-revolution press began to make its mark. Some progress had been made by manufacturers but they had difficulty in controlling the heavy reciprocating bed at the point of reversal, as machines grew bigger and stronger. A Chicago pressman, Robert Miehle, in 1884, solved the problem with a gear which was powerful enough to control the forward motion of a massive bed. The first machine to his design was built in 1887 and in it the cylinder revolved continuously, being raised after impression to clear the forme during the second revolution. The *Miehle*, which had great impressional strength, accurate register and durability, took the lead in the type of machine. With the development of printing from half-tone blocks, followed by process colour printing, such a machine was needed, and it kept its lead for some eighty years.

The Miehle Printing Press and Manufacturing Company, of Chicago, also became well known for a novel automatic stop-cylinder press, the invention of Edward Cheshire, of Milwaukee, in 1920. In effect, this was a flatbed cylinder machine stood on end, with feeding and delivery carried out horizontally. The forme is placed vertically and during impression the cylinder and bed move up and down. Cheshire built four machines and then sold out to the Miehle Company, the first *Miehle Vertical* being built in October 1921. It was to prove very popular in many parts of the world. The printing surface was $12\frac{1}{2} \times 9$ inches, and speed about 3,600 impressions an hour. In 1931 improvements were made and the speed was increased to 4,500 i.p.h. By 1940 further improvements brought production up to 5,000 i.p.h., and further modernization took place in 1964, and this advanced design continues to be manufactured under the name of the *V50 X Extra-Vert*.

The *Miehle Vertical* was manufactured in Britain from 1952 until 1968 at the Goss works in Preston. During this period the Holmes Engineering Company of Birmingham made a copy of an earlier version, which they called the *British Vertical* press, and a limited number of presses was sold.

To describe the numerous cylinder machines made over the years by scores of manufacturers would be tedious, but some note should be taken of particular developments. Printing on both sides of the sheet, and in more than one colour made progress in this period, but apparatus for automatic feeding of sheets and taking them off was somewhat slower in development. The cylinder machine was responsible for a change in the labour structure of the printing trade, a change speeded up by the arrival of the rotary press, although the pressman (renamed machine manager, or minder, in Britain, but retaining his name in the United States) continued to exercise the 'mystery' or art of printing. Problems unthought of in hand printing had to be tackled and the use of power very gradually began the transformation of a trade into an industry.

Early perfecting machines developed by such firms as Middleton, and Dryden and Foord, consisted of two large cylinders, inner and outer, having intermediate drums between for the turning and reversing of the sheets, but the coming of the *Anglo-French* machine did away with the intermediate drums. The *Anglo-French* machine, as its name implies, was the result of collaboration between Marinoni, of Paris, and Hopkinson and Cope, of London. It was based on a Napier design which had been altered in France.

It differed from the earlier machines in that it possessed but two cylinders, the sheet passing directly from the inner to the outer forme. Nearly all the surface of the cylinders was used for impression, there being no idle space under which the forme could return. It was therefore necessary that the cylinders should be alternately lifted from the level of the type in order that the forme could travel back and clear of the cylinder. This was affected by means of a rocking frame in conjunction with knuckle joints or levers.

A separate apparatus for feeding a sheet of paper, called the 'set-off sheet', in with each sheet to be printed was provided. The sheet met this 'set-off sheet' as it entered the second cylinder, and, passing round, with it, prevented the ink on the printed side of the paper 'setting-off' on the cylinder, and from it on to the following sheets. Each machine was arranged with four inking rollers to each forme, making the machine available for the best woodcut work. The speed claimed by Hopkinson and Cope was 1,000 an hour, but if the 'set-off sheet' apparatus was used it was advisable to work more slowly at about 800 to 900 impressions an hour.

Another printing manager of a periodical, Joseph Pardoe, approached B. W. Davis to make a perfector for printing the *Queen*. The machine, named after the periodical, was something like the *Anglo-French*, but differed in certain respects. The two small impression cylinders worked in gear with the table racks and by a motion on either side, one cylinder wheel was alternately thrown out of gear as the tables reversed. When the outer forme was being printed, the cog-wheel on the cylinder was in gear with the rack, the corresponding wheel on the inner cylinder on the offside ran loose on the shaft. Immediately the tables were reversed to allow the cylinders to travel in the same direction, the outer cylinder ran freely back in the rack loose on its shaft, while the other side, locked, drove both cylinders.

Dawson made a perfector on the lines of the earlier models, but which had certain improvements, including a taking-off apparatus. By a series of tapes the sheet was conducted from the outer-forme cylinder across the space the taking off used to occupy, and passed over a first register drum, back round a smaller roller over the second register drum on to a flyer, which carried it over a sloping board immediately in front of the laying-on board, depositing it in full view of the taker-off boy.

Payne, Dawson's former partner, also made a perfector, using large impression cylinders, but bringing them close together, and working in gear, the cog-wheels being comparatively small and compact. Newsum's, of Leeds, copied the *Queen* machine to some extent. The Newsum machine's most novel feature, however, was that it could be used either to print two distinct formes on one side or on both sides of a sheet. Two laying-on and taking-off boards were provided, and by throwing a centre cam out of gear, the cylinders became independent of one another, and delivered the sheets on to their respective taking-off boards.

71. A two-colour Koenig & Bauer machine of 1889

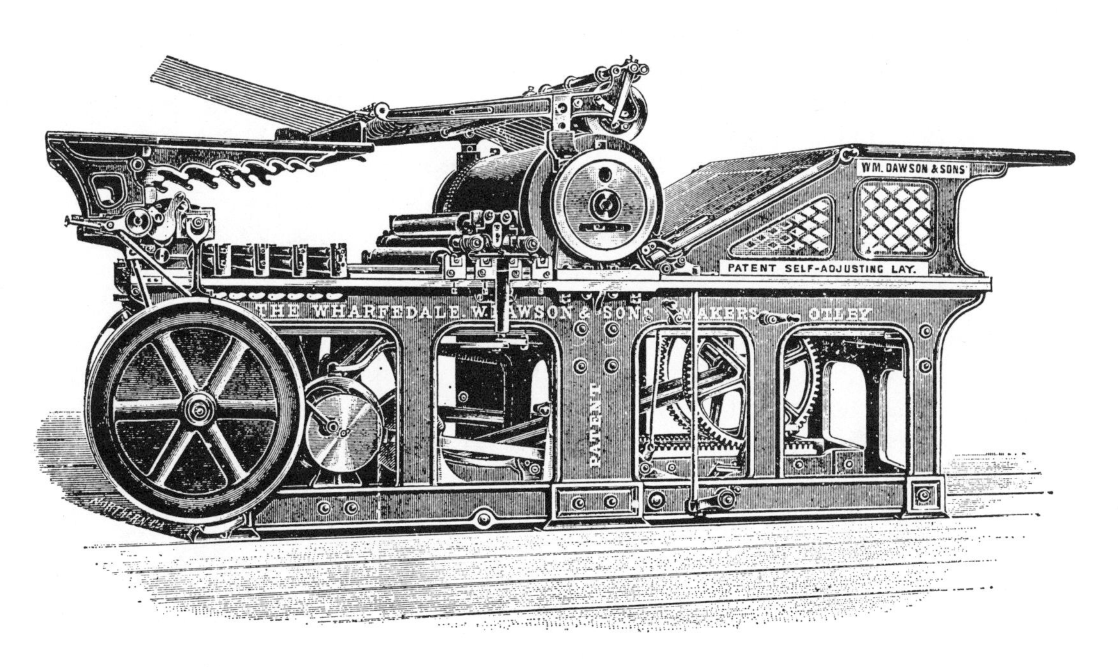

72. A later Wharfedale (1898) by William Dawson, with patent self-adjusting lay

Buxton, Braithwaite and Smith produced a single-cylinder perfecting machine. The cylinder was provided with a double set of grippers, one set being placed in the ordinary position at the surface of the laying-on board, and the other half-way round, slightly over the extreme top. The surfaces between the grippers were each equal to the size of the forme, so that both inner and outer formes had their special impression surface—in the same way as two cylinders on the normal perfector. The sheet was laid in the ordinary manner to the first set of grippers and the impression was given by the inner forme. The sheet then travelled half-way round the cylinder, when it was released into a set of tapes fixed into the supplementary drum on the top. Thus the sheet was reversed—the printed side turned downwards, and run on to the second laying-on board.

When this movement took place the front of the laying-on board was necessarily a few inches above the top surface of the impression cylinder. Immediately, however, the sheet was wholly deposited on the board, the front end of the latter dropped in order to bring the sheet exactly in position for the grippers on the impression cylinder to grasp it.

Marinoni made a machine, like that of Newsum, which could be used either for perfecting or for the printing from two distinct formes. The main movements were based on the *Anglo-French*, but in place of the taking-off board a second laying-on board was substituted, the sheets were carried, after printing, by a series of tapes to the extremity of the machine, down on to flyers, and duly deposited. When it was desired to perfect by altering the gripper shape and adjusting the set-off apparatus the machine was converted to one comparable to the *Anglo-French*.

To consider the question of printing in more than one colour it is desirable to go back to the beginning. The first productions of the printing press followed the style of the manuscript book, and additional colours—illumination and rubrication—to the basic black were added by hand. But at a very early point two pioneers, Johann Fust and Peter Schoeffer, attempted to print in more than one colour with a single pull. In their 1457 *Psalter* there are a series of red and blue initial letters, examination of which shows them to have been printed and not painted in by hand; and the perfect register of which indicates simultaneous printing.

Multiple inking of a single forme is possible when there are broad areas of colour, but attempts at inking individual letters in different colours have rarely been successful. Inking a forme with black ink and then wiping some letters and dabbing them with red ink in order to print at one pull is usually betrayed by a lack of cleanliness at the edge of the red letters. The Fust and Schoeffer initials are so clean and precise that the conclusion has been reached that each initial was made up of a basic relief block with each element to be coloured engraved on a separate thin metal plate which could be inked separately and dropped into its place in the surface of the block.

It is also conjectured that Fust and Schoeffer abandoned this advanced form of colour printing in about 1460 possibly because of the excessive time and money spent on it, and went back to hand rubrication.

Their idea was revived some 350 years later by J. H. Ibbetson and by Sir William Congreve, Ibbetson hinting that Congreve had copied him. The technique was to divide a relief block into a number of parts, inking each with a separate colour and

73. Harrild's two-colour Bremner machine

fitting them together for single-pull printing. It was a solution put forward to solve the problem of bank-note forgery, and as Congreve was a member of the Royal Commission which investigated Ibbetson's idea among others, he may well have derived his idea of compound plate printing from it. Whether this was so or not, Congreve patented a system and machine between 1819–20, for printing the backgrounds of bank-notes in coloured patterns.

A plate made of two kinds of metal and engraved with a design for relief printing was capable of being separated, inked with different coloured inks and brought together again in one plane for printing at a single impression. Ibbetson was hardly a pioneer of this technique, which, as has been pointed out, is nearly as old as letterpress printing itself, but Congreve went further and produced a machine in which the surfaces descended to different levels to be passed over by inking rollers connected with separate ink ducts and which united on rising to form the single printing surface. This system and machine were used to print country bank-notes, paper duty labels and similar 'security' printing, and continued in use for the printing of Government seals until as late as the 1920s.

Charles Knight, a popular publisher, patented his relief colour process in 1838. He used a wood engraved key block and added additional colours by the use of metal blocks. For this purpose Knight envisaged a number of presses. The first was for printing four colours in succession. The blocks were fixed on four beds hinged to the sides of a square table which were turned backwards to be coloured and downwards for the impression. This was provided by the rising of the table caused 'by any ordinary system of levers and screws'. Each bed with its attachments was counterbalanced. The second was a rising table, carrying tympan and frisket, acting in succession against the printing surfaces fixed to the sides of an octagonal prism, the inking being performed by hand at the top of the prism.

Knight wrote that this prism might be used in combination with a *Ruthven* press made deep enough to receive the revolving prism, so that its upper surface could be held in the situation normally occupied by the bed. His third press was an adaptation of the common press. On the normal bed was a secondary bed, turned by a key, and held by a catch in four different positions (somewhat like a railway turntable). The printing surfaces were fixed to four 'tertiary' beds or discs, and revolved on the secondary bed by the revolution of which they were successively brought over four different sheets on the ordinary or primary bed. A centre wheel was so arranged that the tertiary beds revolved so as to be in the same relative position to the different parts of the press.

It will be noted that Knight did not envisage printing in more than one colour by a single impression, but he did print one colour on top of another, while the ink was still wet, thus anticipating a modern technique, known as 'wet-on-wet' printing.

For his well-known system of colour block printing George Baxter (1804–67) used a *Stanhope* and a *Cogger* press as he was convinced that the vibration of the cylinder machine would prevent the accurate superimposition of each block in the process. He made an engraving of the *Paragon* press for his father-in-law, Robert Harrild (see Chapter 5), but there is no record of his using this press himself.

Printing in several colours simultaneously is possible by using a split ink duct—

that is, one with separate supplies of ink and an inking apparatus divided accordingly. From 1844 onwards there was a stream of patents for simultaneous colour printing, particularly in the United States—proposing the most diverse techniques, ranging from ink troughs with sliding partitions to revolving platens. A typical invention was the Chromatic printing press made in 1871 by Suitterlin, Claussen & Co., of Chicago and New York, which printed three colours from one forme at one impression. The surface of the inking cylinder was divided into three equal parts, and supplied with adjustable sectors (or colour stripes) of various sizes to correspond with any line or part of a line of type. Each part was supplied with a colour from one of the distributing rollers. Basically the press was of the job platen kind with inking rollers. A speed of from 1,000 to 2,000 impressions an hour was claimed, depending on whether it was worked by treadle or steam-power.

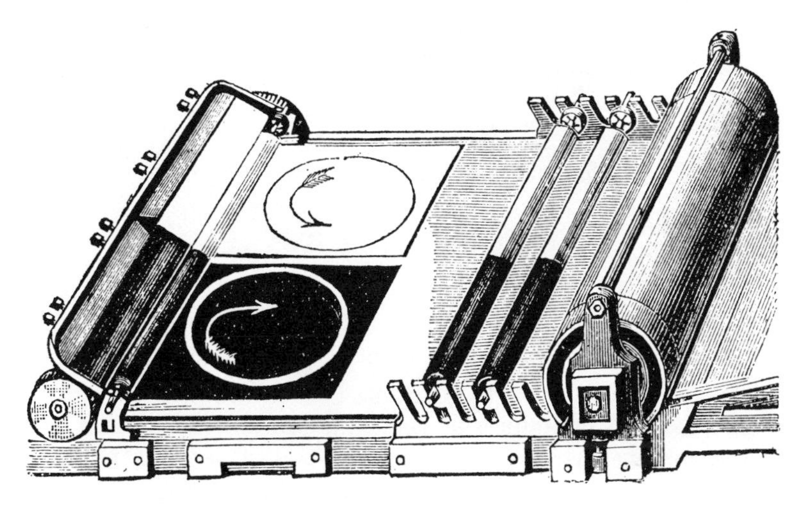

74. Rutley's two-colour inking table

In Britain there was, for example, Rutley's two-ink colour table of about 1879. This device enabled a printer to print in two colours at once on a jobbing platen—there were two inking discs which fed different colours to each half of the rollers.

Fountain dividers became common and pressmen at first tended to cut grooves in the rollers to prevent the colours mixing together, but separating devices were invented to eliminate the need to cut the rollers.

Other presses did not, strictly speaking, provide simultaneous printing in various colours, despite their claims to do so. They were ingenious devices by which colours were applied by separate printing surfaces but in a continuous operation. The *Chromatic* press, developed by A. A. Dunk, of Philadelphia in 1871, was said to print five or more colours at the same time. It consisted of two sectional cylinders revolving in unison, one carrying formes and the other a corresponding number of tympans. A skeleton cylinder contained the nippers, by which the sheet was retained until fully printed. The sheet was presented to the successive formes, from each of which it received an impression in a different colour.

75. The Leeds two-colour machine

In Britain an ambitious attempt was made by Newsum, Wood and Dyson, of Leeds. Their *Leeds* two-colour 'rotary machine' consisted of two cylinders working together, the smaller (or impression) cylinder making two revolutions to one of the larger. This was of an unusual construction, two areas being flattened, on which the formes for each colour could be fastened. Duplicate inking arrangements were incorporated with rollers for each colour, which avoided each other when in operation. A machine was installed at Mr. Cowell's establishment in Ipswich, and was so great a novelty that printers journeyed there to see it at work. Nevertheless, it does not seem to have become very popular any more than a number of similar devices. In 1909 the Miehle Company added the *Upham* Colour attachment to its flatbed cylinders, which allowed curved plates to be mounted and printed against the impression cylinder, thus adding a second colour after the first printing had been made from the flat forme. The mechanism was not widely adopted. In the 1960s the Heidelberg Company revived a similar idea with its combination flatbed and rotary machine.

In general, printers have continued to keep to the method of printing each colour separately. Moxon explains how it is done on the hand press. A forme of type is made up and a frisket cut for it. The compositor then takes out the type to be printed in, say, red, filling up the space with quadrats and spaces. The pressman works off the black impression, and following this the quadrats are removed from the forme and slips of board, called underlays, are put in their place. The types to be printed red are placed on the underlays, so they stand higher than the rest of the matter in the forme and will print where the rest will not. A new frisket is prepared to mask everything but the 'red' types, and the black-printed sheets are then printed with the red ink. Later manuals suggest printing the red before the black, but the basic point is that in letterpress printing each colour requires a separate impression.

This could be a slow process, particularly if a printer owned only one press. J. M. B. Papillon, a wood-engraver, in his *Traité historique et pratique de la gravure en bois* (Paris, 1766), does refer to George Lallerman's 1623 invention—a press which had three platens side by side, all operated by one pull of the bar. As can be imagined, the results were not good and so Lallerman abandoned this approach in favour of a press in the style of a copperplate press. This had three tables and six rollers, and required four men to work it. Each of three blocks was inked by a different man, while the fourth turned a handle to produce three impressions simultaneously. Both Lallerman's presses seem to have been commercial failures.

While printing was at the hand-press stage not much could be done to speed up colour printing, but the coming of the cylinder machine opened up the possibility of coupling single-colour printing units together—for example, two printing cylinders and an intermediate one to carry the sheet from one to the other to receive a second printing. This, in turn, necessarily led to devices to speed up drying, such as gas jets, or, later, electrical heaters.

As has been mentioned, however, Charles Knight pioneered 'wet-on-wet' printing as early as 1838, and as practised today this consists of printing in two or more colours where the ink film of the first colour is still wet when the paper receives a second or subsequent impression. The consistency of the inks must be progressively less tacky, and special wet-printing process inks are made for use in multi-colour printing. They are quick-drying and the ink of the last colour to be printed binds the others together.

The first really successful two-colour machine was made by Conisbee and was bought by the firm of Waterlow in 1861. Similar to an ordinary single-cylinder machine based on the *Main* principle it was provided with two sets of inking apparatus, each of which acted independently of the other. The cylinder was placed in the centre and made two continuous revolutions, giving an impression for each colour. Conisbee introduced a perforated gas pipe under each end so that the ink tables could be gently heated to assist even distribution of the coloured inks. The machine was not very fast in operation, the rate being only about 300 to 400 copies an hour.

Conisbee was followed by Dawson, Harrild and Payne who all produced a single-cylinder two-colour machine. Dawson erected for the firm of Wilkinson, of Pendleton, a *Wharfedale* capable of taking a sheet 75 × 60 inches, by far the largest machine ever attempted up to that point, and which printed 600 impressions an hour.

The idea of having two cylinders for colour printing appears to have been first employed by the American manufacturers, Huber and Hodgman, of Taunton, Massachusetts, in the 1880s. The sheet was fed on to the first cylinder to receive the impression from the first forme, and was taken by grippers to the second cylinder and carried up between the two cylinders and the bed again, where it received an impression from the second forme.

The arrival of the cylinder machine not only gave rise to a new class of pressman but to others who fed and took away the printed sheets. Originally, pressmen and their apprentices—those destined to become pressmen—performed all the tasks necessary to printing, but with the development of the newspaper in the eighteenth century and the consequent need for greater speeds, the traditional labour pattern began to change. An unindentured boy might be employed to fetch and carry and to act as a 'fly' to take the sheet off the tympan as the pressman turned it up. He was the forerunner of the 'taker-off', who received the sheets from the cylinder machine and placed them on the heap.

The pressman was transformed in Britain into the machine manager, or minder, and carried on with the mysteries of make-ready, but ceased to be greatly concerned with feeding the sheets of paper and carrying them away. So careful was he of his specialized knowledge that he might even have placed a screen round his machine while he was doing his make-ready. This consisted, basically, of equalizing the impression, perhaps by means of overlaying, or placing pieces of paper on the cylinder to press the paper down in those parts of the forme which were lower than others; or underlaying—the placing of sheets of paper under the forme.

The preparation of the cylinder was also of great importance. The object was to place around the cylinder a foundation on which the sheets could be fastened and which would act against the type in the manner of the old tympan. Calico was used as a base with sheets of hard paper on top, the number being determined by the size of the forme to be printed. Much skill was needed in pasting the sheets together and putting them round the cylinder, adding the 'blanket' (made of cloth or even of rubber), and the preparing of the overlays, particularly if the machine was old, the type worn and the printing surfaces uneven. Hard-packing—dispensing with the blanket—was introduced in the 1880s to Britain from the United States, but there were always different views as to its efficacy.

The machine minder, who was responsible for the machine, the inking and good impression, proved to be long enduring, but the boys, and later the men and women, who fed the paper and took it away were eventually replaced by machinery, although it took some time. At a quite early date in the life of the cylinder machine, inventors were trying to replace this human element, towards whom printers had an ambivalent attitude. On the one hand they were cheap, but on the other they were unindentured labour and were a nuisance. One writer (in the *American Dictionary of Printing*, 1894) put his finger on the problem when he wrote: '. . . the certainty that they cannot rise operates to keep them unsteady'. But they stayed as long as the inventors failed to produce a satisfactory substitute.

When Edwin Norris in 1835 took out a patent for a printing machine (see Chapter 9) he also included a feeding and taking-off apparatus. 'At each end of the press are jointed tympan frames of a peculiar construction, which are periodically depressed by the action of a cam so as to come down over the cylinder, and (by the winding of a tympan cloth which recovers itself by a spring) feed in a sheet to receiving rollers near the top of the cylinder', he wrote, describing one aspect. The sheet was discharged on to a 'depositor' (a travelling frame with endless bands), by which it was laid gently upon a receiving table at the end of the press. Andrew Smith, in the same year, in the course of a long patent, describes fingers or 'gripers', as he calls them, which perform a variety of functions including feeding and taking off, but little more is heard of them.

The 'flyer' mechanism, as developed by Hoe and others from 1830 onwards, consisted of a small drum on the top of the impression cylinder and working in gear. It was provided with a set of grippers so arranged as to enter the opening of the former immediately the sheet was released, at the same time clutching it and releasing it on to a series of endless tapes. From the tapes the sheet ran round a small set of upper wheels and then fell to a lower set in front of the 'flies'—wooden laths originally—in the form of a comb. These worked on a rod axis and alternately assumed a perpendicular and horizontal position. The sheet clung to the flies while in the process of falling and when horizontal laid in a heap to be taken away. Ringwalt wrote in 1871 in his *Dictionary* 'Nearly all the superior class of machines are now furnished with flies, as they effect an important saving of labor'.

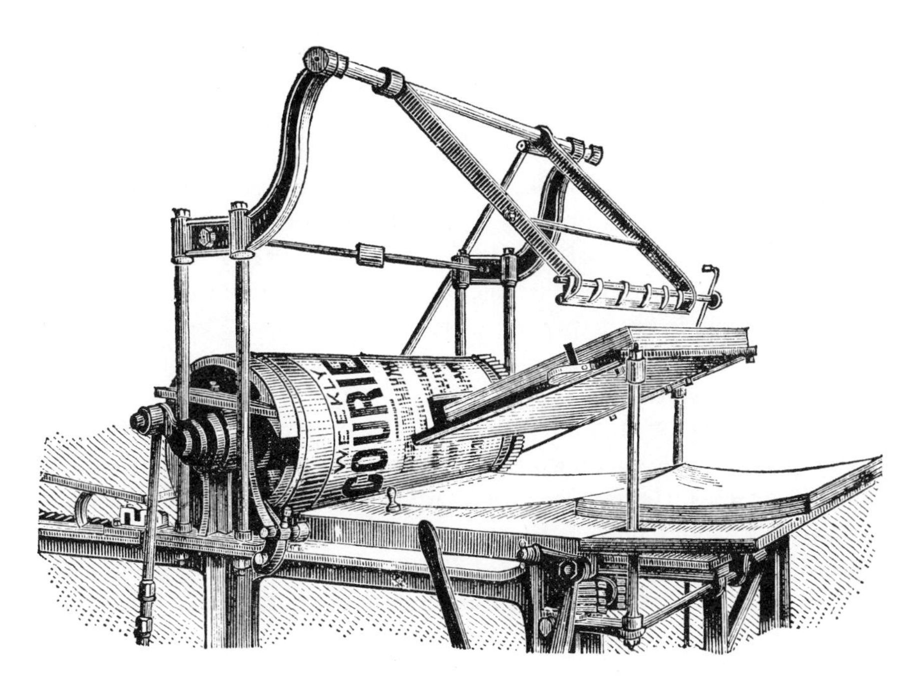

76. Mark Smith's taking-off apparatus, *c.* 1879

Mark Smith's taking-off apparatus, introduced in 1878, consisted of four uprights fixed to the side frame, immediately above the cylinder. On either side of the cross-bars of the uprights an arm was attached with a second frame dipping down towards the laying-on board, having a series of grippers arranged along a bar at the extremity. The frame had a movement extending from the base of the cylinder on the laying-on board to the taking-off board, and was supported and moved by a steel bar extending from a wheel on the cam shaft outside the side-frame. When the machine was set in motion the frame moved downwards and the grippers entered the cylinder, opening immediately upon the release of the sheet, grasping it and returning above the taking-off board, dropping the paper on the heap before again moving downwards.

Eventually, taking-off apparatus developed to become an extension delivery and automatic pile-lowering mechanism. The extended delivery—an addition to the front of the press—which seems to have been introduced in about 1913—eliminated what was known as slip-sheeting. A slip-sheet is an extra sheet of waste paper placed between freshly printed sheets as they are printed to prevent set-off of ink. The slight time lag involved in getting the sheet to the extension assisted in the drying of the ink.

The 'taker-off' departed from the industrial scene well in advance of the feeder, although inventors were continually trying to produce a mechanical alternative. On behalf of the Hoe firm, William Newton, in 1844, patented a press which had a 'peculiar' arrangement to raise the feed table, and a further contrivance which caused spring clips on the cylinder to open and seize on the sheet 'which is by the further rotation of the cylinder wound round it and thus carried to the impression'. This does not sound as if it was the right approach. Nearer the modern method was that of George W. Taylor, of New York, in which a number of vertical cylinders with pistons were placed above the sheets. A revolving crank shaft moved the pistons and made air pumps, forcing the air through the tubes as the pistons rose and drawing air through as it descended. The piston rods, in sliding on the paper, moved the top sheet towards the ends of the tubes. As soon as the top sheet arrived at the tubes, the ingress of air was checked, causing a partial vacuum to be formed in the cylinders above the pistons, the pressure of air below preventing the piston rods working up in the paper until the sheet was carried away. One of these devices was used in the printing shop of the *Christian Advocate and Journal* in New York on an A. B. Taylor press.

That was in 1857, but in 1853 John P. Comly, of Dayton, Ohio, had patented a feeding apparatus which worked by means of atmospheric pressure, using exhaust pump and tubes. Henry Clark, of New Orleans, the next year resorted to the more primitive 'pushing motion'; George Little, of Utica, New York, to indiarubber guides; and William F. Collier, of Worcester, Massachusetts, to a bellows. Every kind of idea was tried out—vibration, currents of air, 'fingers' and guide tapes. In the ten years from 1853 at least twenty inventions were registered in the United States alone, but little advance was made.

By June 1890 the *Printers' Register*, in London, was led to comment that for the last forty years the attention of inventors had been directed to the rendering of an ordinary single-sheet printing machine altogether automatic, in which quest immense sums had been lost. Experience had shown that the inventions were unreliable and those who were interested in automatic feeding had fallen back on feeding from endless rolls of paper.

However, the *Printers' Register* was able to report on the new invention of Edward T. Cleathero, an engineer, and Joseph A. Nichols, a printer, which had been patented the year before. It may be a coincidence, but 1889 was the year in which the layers-on had revolted against poor conditions in London and had formed their own union. The Cleathero and Nichols feeder consisted of a vertical sliding frame carrying a horizontal spindle with indiarubber discs which rested on the pile of paper so that when they revolved they fed the top sheet forward from the pile. A piece of each disc was cut off so while performing a complete revolution they came to rest temporarily when the flat part was presented to the pile of paper. The sheets moved forward and were seized by gripping rollers, then, in remaining stationary, the discs allowed the remainder of the sheet to be drawn from under them. The period of rest was made to correspond with that of the feeding cylinder so that the sheets were fed forward exactly in accordance with the motion of the cylinder.

But once again the significant advance was in the United States, where in 1897 Talbot C. Dexter developed an automatic paper sheet feeder which separated individual sheets from the top of the pile, fed them down a conventional feedboard and registered them to the press guides. Frank L. Cross in 1905 employed a combing action for separation of the sheets, which were placed on a top loading board, where they were moved backward around a drum and then forward on to the feedboard by the combing action.

From the turn of the century four types of feeder have been developed on different principles—suction friction rollers, combing and corner separation—using blasts of air and suction feet. By 1914 the *Kelly* press, built by the American Type Founders Company, was the first of the flatbed cylinder presses to be designed with an integrated feeder, employing a combination blast and suction device.

Finally, therefore, the cylinder machine became almost automatic in operation, but it brought with it problems the hand pressman never had to face and also possibilities which he barely dreamt of. Static electricity caused the sheet to cling to metal and wood, and since 1910 static eliminators have become available. The various problems arising from set-off, which were tackled in the early days with 'set-off' sheets and 'slip-sheets' have since about 1904 been counteracted with anti-setoff sprays and by inks which include special ingredients.

Steam early replaced manual power but continued in use for many years. Steam-engines continued to supply the power for printing *The Times* as late as 1908, although the first electrically driven press was installed in 1905. The gas-engine was adopted from about 1860 onwards, but did not really come into its own for another twenty years, by which time printers were thinking about electricity.

An early attempt in Britain at running a printing machine by an electric motor was made in June 1884 by the *Somerset County Gazette*, but a few months earlier, in March of the same year, in Ilion, New York, the *Ilion Citizen* claimed that it was the first newspaper in the United States to be printed with the aid of an electric motor.

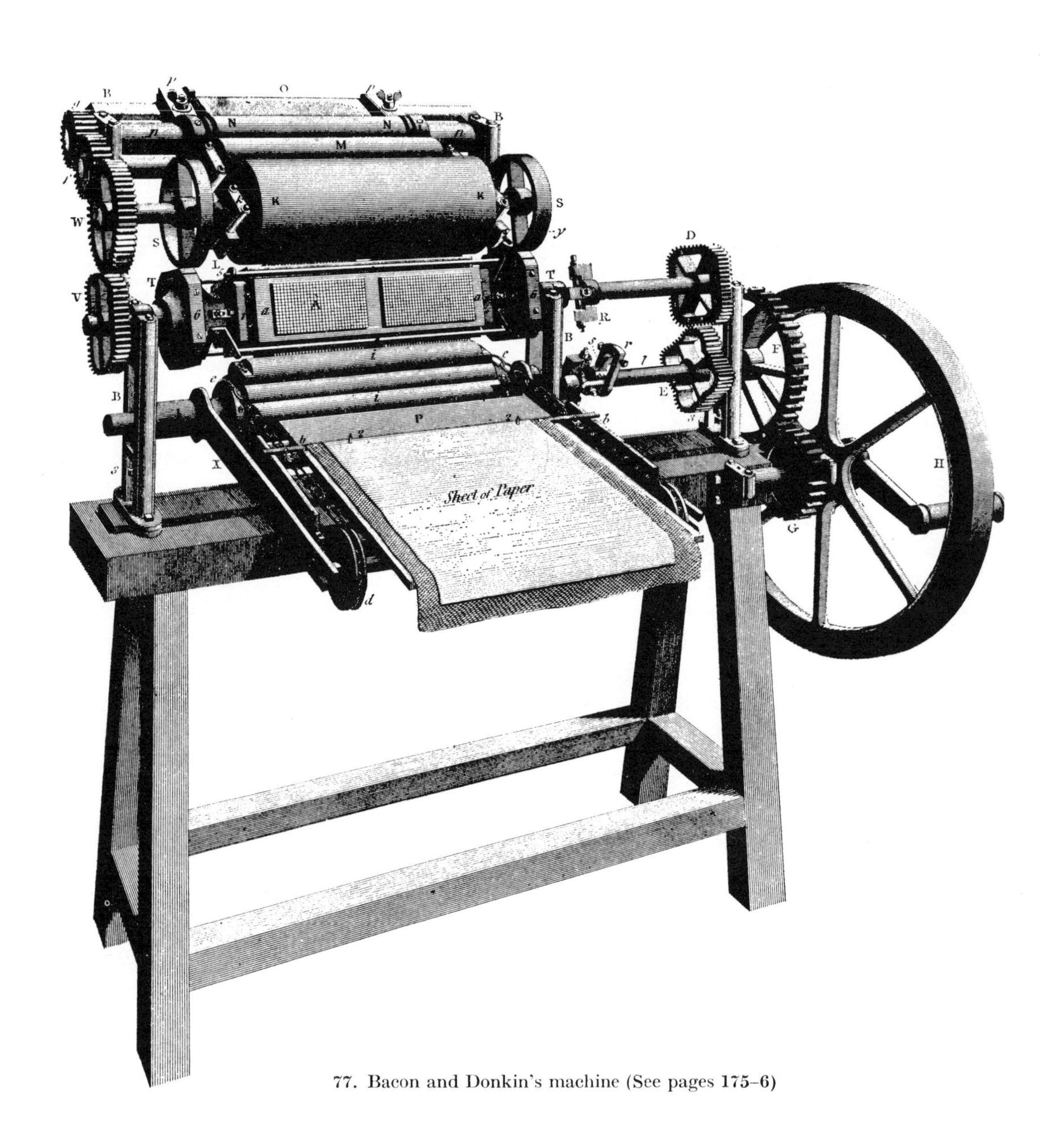

77. Bacon and Donkin's machine (See pages 175–6)

12

ROTARY PRINTING—THE BACKGROUND

THE ORIGINS OF the rotary method of printing reach back at least two hundred years and much debate among nineteenth-century inventors and their supporters regarding priorities could have been avoided if they had conducted some research into the past. Rotary printing, in which the material to be printed is fed between revolving cylinders, one of which carries the printing surface, derives not from relief-printing techniques but from the intaglio process, in which engraved copper plates are printed from a rolling press. A layer of felt and a sheet of paper are placed on top of an inked engraved plate, which is then rolled between two cylinders in the manner of a mangle. It is not certain when the rolling press was invented, but it was in use in the sixteenth century. George Tomlyn, in his patents of 1660 and 1662, for a new way to ornament vellum and parchment, uses the words 'with a rolling printing presse and ingraven plates', indicating that by that time the device was commonplace.

If flat engraved plates could be printed from in this way, there was no reason why one of the cylinders could not be engraved, and reels of material fed between it and the pressing cylinder. This idea certainly proved attractive to printers of cloth in the eighteenth century, and experiments might even have been made in the previous century, but this is difficult to ascertain.

Unfortunately, early English patents do not provide details of the devices they were issued to protect, tending to fall back on such vague phrases as 'peculiar manner'. This is used in Patent No. 71, of 1634, issued to Arnold Rotsipen (or Rotispen) for a 'pressinge or printinge engine with wheels & rolls after his peculiar manner'. Patent No. 304, of 1692, issued to William Bayly, is not much more helpful. It is for 'A new art or invention for printing all sorts of paper of all sorts of figures and colours whatsoever, with severall engines made of brasse and such other like mettalls, with fire, without any paint or staine, which will be usefull for hanging of rooms, and such like vses.' Whatever else it was, Bayly's machine was meant for printing cloth and if paper was to be printed it was not for reading purposes.

A hundred years later patents were more specific, and in 1764 Thomas Fryer, Thomas Greenough and John Newbery patented a machine 'for printing, staining and colouring of silks, stuffs, linens, cottons, leather and paper'. The printing was to be performed 'by means of engraved copper cylinders on which the colours are laid by smaller cylinders which are put in motion by other plain cylinders, and the whole work of filling in, clearing off, and stamping the impression is performed by the joynt assistance of sundry springs and the intermediums of coggs and rings turned by a wheel worked by either a horse, water or wind'. Here we are in the presence of true rotary printing.

In 1772 Joseph Adkin, the elder, Joseph Adkin, the younger, Charles Taylor and Thomas Walker patented 'a new machine or engine for stamping and printing of paper, silk, woollen, cotton and linnen cloths, and other articles of silk, wool, cotton. . .'. The press was composed of three cylinders arranged vertically; the upper one acted as a pressing cylinder; the middle one, to which the power was applied, as a printing roller; and the lower one as a furnishing roller, it being partly immersed in a colour trough. It was calculated that the press could print a fabric 22 inches wide. The fabric passed to the impression 'through two rails fixed upon the sides of the uprights upon the level with the middle roller, to prevent the cloth from crumping'.

Eleven years later, in 1783, and again the next year, Thomas Bell, a copperplate printer, took out patents for a printing machine to print up to five colours on 'linnens'. Otto M. Lilien, in his *History of Industrial Gravure Printing up to 1900*, London, 1957, comments: 'The patent drawing and description show a machine with all the essential parts of a present day industrial gravure machine, whether used for printing on paper, fabrics, or plastics, and explain in detail the ink duct, the cylindrical printing forme, the doctor blade for wiping the surplus ink from the cylinder and the impression roller.'

In 1786 Jacob Bunnett patented 'a machine for printing of paper hangings, calicos, cottons and linens in general, whereby any number of colours may be printed thereon at one and the same time, and whereby ten times as many pieces may be printed in as short a space of time as one piece is not printed by the common method.' Three horizontal and parallel 'printing cylinders' each had a waterproof colouring roller above them and on the top of each colouring roller a canvas or wire-bottomed trough. The pressing cylinders were below the printing cylinders and were thrown in and out of gear by the action of wedges on their axles. The material passed before being printed over six rollers 'to give it friction and to prevent its wrinkling'.

In France, J.-A. Bouvallet, of Amiens, after 1770 made a machine which utilized an engraved copper cylinder for printing wool and plush; and the famous Swiss textile printer, Christophe Philippe Oberkampf, who set up at Jouy, near Versailles, had a rotary press for printing calico in 1780.

The development of rotary techniques for the printing of textiles was sufficiently advanced by the 1830s for a subsidiary supply industry to have grown up. Richard Ormerod, of the St. George's Foundry, London Road, Manchester, in his 1833 price list is pleased to announce himself as sole agent for Vivian & Sons improved patent refined copper rollers for calico printers.

These calico printers were clearly years ahead of either Nicholson or Apollos Kinsley (who has been referred to in Chapter 7). When Nicholson, in 1790, took out his patent, which is often hailed as the first mention of rotary printing, he had had the opportunity, as a patent agent of sorts, of studying former patents for presses utilizing the rotary principle. Theodor Göbel, the nineteenth-century apologist for

Koenig, was of the opinion that Nicholson obtained his idea from the 1772 patent of Joseph Adkin and partners.

In any case, Nicholson's patent was never put to the test, any more than was that of William Church (1823). Church showed remarkable foresight with his idea of types locked round a cylinder; colour printing by three cylinders arranged around a pressing cylinder; and heated tubes to dry the fabric when passing from one impression to another. His 'friction lever' to affect tension on a roll of material is particularly ahead of its time, but there is no evidence that any of his mechanisms was ever built, or, at least, installed in a printing office. Church is well known as perhaps the earliest inventor of a composing machine, but, again, it is not known whether anything but a prototype was built. Church was a visionary, but he does not deserve the scorn with which he is treated by Hansard.

Another half a century was to pass after Nicholson's patent before relief rotary printing from reels of paper became a commercial proposition, but during this period there were many experiments, and it is difficult to apportion credit to the various claimants to the title of 'inventor' of the rotary printing press, if, indeed, any one inventor can be picked out.

Before relief rotary printing could become a reality, the problem of fixing type round a curved surface had to be solved—a more complicated task than engraving on the round. Nicholson had suggested types which were to be scraped to make the tail smaller than the top, so that they could be 'firmly imposed upon a cylindrical surface in the same manner as common letter is imposed upon a flat stone'. These types were to be imposed in chases adapted to the surface of the cylinder and fastened by screws or wedges, or in grooves. Nicholson was something of a theoretical pioneer in this respect, because, apart from his primitive notion of 'scraping', specially cast types were eventually fitted around a cylinder as he suggested. But this was only an interim solution, since the real approach was to cast a curved stereotype which could be fitted round the cylinder. The way to do this was not immediately apparent, and Cowper's method, introduced for his security press in 1816, involving a laborious heating process, could not have appealed to newspaper printers.

Another obstacle in the way of the wider adoption of rotary printing related to the feeding of paper. If the time-consuming feeding of individual sheets by hand was to be avoided then a continuous feed from a reel of paper was the obvious way out. Here the barrier was not technical put political, since the British paper tax required that individual sheets of paper had to be stamped. Until the tax was repealed sheet-fed machines continued to be used for the most advanced newspaper printing, and the desire to increase productivity led to an increase in the number of feeding stations on presses until they became too large and unwieldy to accommodate.

Before the repeal of the paper tax in 1861, the rotary press underwent development in stages. The first to be built, that of Richard Bacon and Bryan Donkin, patented in 1813, was fed by sheets of paper and avoided the problem of fitting type to cylindrical surfaces. The type was still held in flat formes, which were fixed on four sides of a prism, which was square in section. Its axis revolved by the action of a winch, and the type was printed on to the paper by means of a second roller, called by the old name of the platen, its surface being made up of four segments of cylinders, and its circumference when turned round always applying to a type surface. Ink was applied by a large composition cylinder above the prism, which received ink from a distribution roller supplied from a third metal roller. Bacon and Donkin were thus pioneers in the use of the composition roller and the ink duct. The whole mechanism was quite small, capable of standing upon an ordinary writing-table, but it was very complicated and required great accuracy of operation. An exhibition was held in Donkin's factory, and claims were made that the machine would perform the work of eight hand presses. Hansard states that he showed the inventor that work on six of his presses would have required four of the new machines to execute it. The only one of Bacon and Donkin's machines known to Hansard was installed at the University Press, Cambridge, where (in 1825) it 'rests in peace, as not being found in any degree useful'.

The first attempts at feeding from a reel of paper were made in America. The earliest American manufacturers of 'endless paper' were Joshua and Thomas Gilpin, of the Brandywine Paper Mills in Delaware. The paper-making machine had been invented by Nicholas-Louis Robert, of Paris, in 1798, but owing to the unsettled state of France at the time, he sold the patent to St. Leger Didot, whose brother-in-law, John Gamble, travelled to England in 1799 and persuaded the papermakers, H. & S. Foudrinier, to finance the building of the machine. The first papermaking machines were set up at Frogmore, Hertfordshire, in 1803 by Bryan Donkin for the brothers Froudrinier, and at St. Neots, Huntingdonshire, by Gamble. By 1817 the Gilpins had a machine based on English models and as part of the publicity for their endless paper produced a printed work upon it.

They must have constructed some kind of printing machine as the American Philosophical Society noted in its Minute book on March 1819 that they 'exhibited printing on endless paper at the time of its manufacture by a cylindrical water press, and with a degree of rapidity never before equalled'. A reference appears in the Society's Donation book as follows: 'A Specimen of their paper (Endless sheet) printed as it came west from the Paper making rollers—54 Pages Stereotype placed on a Roller revolving in 10 Seconds of time 12mo.'

Whether the press was truly a rotary depends on the interpretation of the phrase '54 Pages Stereotype placed on a Roller'. The machine may have been a cylinder press, but whether this was so or not, printing from endless paper had obviously been accomplished.

On the other hand, the invention of Henry Betts, of Norwalk, Connecticut, patented on 14 September 1833, for 'printing paper on both sides', does seem to have the characteristics of a rotary press. Betts proposed using the whole surface of the cylinder from which to print. Three methods were proposed. In the first, stereotype plates on blocks were hinged together to form an endless chain which was placed round a polygonal roller of wood, strong enough to sustain the pressure of the impression cylinder. In the second, stereotype plates or movable types were to be employed on similar blocks united by strips of iron, and the third method employed two or more movable formes passing over a roller and under a cylinder.

Betts's machine was crude and imperfect and did not get very far, but shortly after he patented his press Thomas Trench, a papermaker, of Ithaca, New York, actually built a web-fed press which is known to have printed books. Trench, the son of a Scottish papermaker, James Trench, who emigrated to America, followed the same line of business as his father, and it was for the purpose of creating a better market for his paper that he invented a machine which would print from a roll of paper. On 20 November 1837 he was granted a patent for a 'machine for printing both sides of a continuous sheet of paper'.

On the two lower cylinders were stereotype plates secured by clasps, screws or 'other suitable fastenings', and at each end were adjustable bearers which could be raised or lowered to allow of a lighter or stronger impression. The two upper or impression cylinders were 'covered with a soft elastic substance for pressing the paper against the types'. The paper was placed between endless tapes 'so adapted in number and position as to fall between the pages of the printing, or on the margin'.

Two books printed on the machine were stated to be an edition of *Robinson Crusoe* and a spelling book. Rollo Silver, who has been responsible for revealing much of the detail of early American web printing, on which this chapter relies, states that it is most probable that the *Robinson Crusoe* is the 1836 Ithaca edition published by Mack, Andrus and Woodruff, which owned the paper-mill in which Trench was employed. The other can be tentatively identified as the 1837 Ithaca edition of *Cobb's Spelling Book*, also published by Mack, Andrus and Woodruff.

At the tenth annual fair of the American Institute in 1837, James Trench received a diploma 'for a sheet of printed paper, about eighty feet in length, containing ten copies of the life of Robinson Crusoe', and in the same year the Mechanics' Institute awarded a silver medal to Thomas for 'specimens of printing from a machine, by which endless sheets are printed on both sides at the same time'.

It seems fairly clear that the first successful web-fed rotary press was built in the United States, although in the decade 1830–40 inventors on both sides of the Atlantic were exploring the possibilities. Indeed, Rowland Hill, pioneer of the penny post, assisted by his brother Edwin, in 1835 invented a rotary machine, which he had built by Dryden of Lambeth (Plate XLIII). He revived the Nicholson idea of printing from movable types fixed around a cylinder. Both proposals called for tapered types, but Hill was in advance of Nicholson's notion of scraped types in that he specified a specially shaped type-mould. Hill's principal suggestion was that of printing from a continuous web of paper, but this was precisely what was not commercially possible while the paper tax remained. In June 1836 Hill asked the Treasury to relax the rule that each sheet be stamped separately but being refused, lost interest in rotary printing and devoted his time thereafter to Post Office reform.

It is not surprising, however, that David Napier, who had been so successful with the cylinder machine, should turn his attention to the rotary press. In his patent of 18 April 1837, he called for part of the 'type cylinder' to be made into a distributing surface for the ink. In fact, the word 'stereotype' appears in brackets after 'type cylinder', but at that point it is unlikely he could have been referring to curved stereotype plates, but rather to some more primitive method as practised by the American pioneers. What Napier was thinking of is not certain, but he produced a machine, if a letter from his son, James Napier, to the *Mechanics' Magazine* of 2 September 1848 is any guide. An account of 'Hoe's Fast Press' had been reprinted in that journal's 26 August issue from the American periodical *Eureka*, and James

Napier was at pains to explain that ten years before the press so described had been made by David Napier, adding: '. . . Mr. Hoe was in London after the patent was obtained, and the invention was then fully explained to him and many details which were not described in the specification, to which, of course, he had free access, were liberally laid before him, by the inventor himself in a friendly way.' James Napier may have been firmly of the opinion that Hoe had filched ideas from his father, but curiously, as head of the firm of Napier's, he did nothing to enter the expanding field of newspaper printing after the rotary began to develop, but turned back to the simpler 'bed and platen' machine (see Chapter 8).

To digress a little, since the name of Hoe at one time became almost synonymous with the rotary press, it can be noted that from the iron hand press to the rotary machine Hoe was accused by inventors of adopting their ideas, and others were of the same opinion, the whole case being put by an anonymous contributor, called 'Franklin', to the *Printer's Circular*. The *Smith* press, he said, was really the *Wells* press; a Hoe double cylinder press of 1832 was 'an exact copy of a British press, to which the Scotch name of *Napier* was attached, and whom I always considered the inventor'. The bed and platen was the invention of Isaac Adams, and the card press that of Mr. Voorhees. The radical feature (viz. a central revolving type cylinder surrounded by a number of impression cylinders) in Hoe's type revolving press was, 'Franklin' claimed, patented by 'Morrison in England six years before Hoe brought out his machine', and the idea of attaching type to the cylinder was procured from a former pressman of the *Tribune* newspaper.

There are at least three points to bear in mind when considering these attacks on Richard Hoe. The first is that a number of inventions were not as original as was thought, as it is hoped has been made clear in this work. A man of the calibre of Napier could not have thought that his patent was complete protection if he explained his machine in such detail to a rival manufacturer. The second is that Hoe genuinely took over patents, and patents were made over to him when the inventor did not have the resources to exploit them; and thirdly, and this 'Franklin' admitted, Hoe had those resources to turn out 'the most carefully finished work'. That is not to say that Hoe was an idealist, and never adopted another man's ideas without recompense, but often those who complained had themselves adapted earlier inventions.

The 'Morrison' referred to by 'Franklin' was David Morison, who took out a patent on 16 December 1839, part of which stated 'when quick printing is required, a stereotype plate is laid on the main cylinder (horizontal), and around this are placed as many pressing cylinders and inking apparatus alternately as the size of the respective rollers will admit of, or rapidity requires'. Nothing is known of this invention, which sounds as if it were too complicated to become a commercial proposition, since it also envisaged turning the centre cylinder into a pressing cylinder and the others into rollers for various colours. At that point, and this applies to Napier's patent also, no real advance had been made in Britain to produce curved stereotyped plates. The first developments in this technique will be outlined, but before that point is reached some mention should be made of other early rotary machines.

The *South-western Sentinel*, of Evansville, Indiana, of 28 February 1840, claimed to be 'the first newspaper probably in the world which was ever printed on a continuous sheet', on a press invented by Josiah Warren, of New Harmony. The press had taken nine years to develop, and the following is a description from the *Sentinel*: 'It receives the paper from a roll, prints it by means of a roller, and winds it as it is printed. It is worked by a man and a boy, or, at somewhat slower speed, by a man alone. It is supplied with self-inking apparatus by which the ink is strictly under control. Its construction throughout is very simple. It has not a single geared wheel about it. It is chiefly composed of rollers, twenty-three in number, with several pulleys. Its form is elegant and its appearance substantial.' However, due to breakdowns and malicious damage, Warren took the press away and broke it up.

The story of Jeptha A. Wilkinson's rotary-press patent and its piracy by Moses S. Beach is long and involved, but Wilkinson's successful memorial to Congress in 1856 for an extension of his patent rights places him in the line of rotary-press pioneers. He had apparently been working on the idea since 1818, but it was not until the period 1837–9 in Providence, Rhode Island, that he made a model printing press, the cylinders of which were about 9½ inches in diameter and printed a width of about 5 inches. A newspaper sheet, entitled *The Endless Register*, dated at Providence, 11 May 1840, was printed. Like other inventors Wilkinson needed funds to exploit his invention and as Moses Y. Beach, proprietor of the New York *Sun* 'held out good inducements' Wilkinson left Providence in March 1841 for New York. Arrangements were made with Beach to construct an entirely new press, and after some delay it was completed so as to print on both sides of the paper, fold, cut and deliver the sheets at the rate of some 300 a minute.

But Beach cut off payments, demanded that the patents be in his name, and sent his son, Moses Sperry Beach, to England to patent the press there. Moses S. Beach was, in fact, granted a patent in London in 1842, which utilized the customary words 'communicated to me by a certain foreigner resident abroad', indicating that he was not claiming the invention as his own. The patent was for a machine with four cylinders, upon two of which the type was 'adjusted'. Damped paper was wound round a shaft, and passed from it between the first pair of cylinders, receiving the impression on the upper side. It then passed between the second pair, receiving the impression on the under side. Inking apparatus was attached to each of the type cylinders (this was not claimed as part of the invention). Lifting rolls or wheels raised the paper to folders, which folded it in its course to the rotary knife. The machine was driven by a crank handle.

The adjusting of the specially cast type on to the cylinders was an elaborate business. Tapered type with projections and indentations could be fitted together in a composing stick with a curved base. When the stick was full the type was taken out and placed in a ring or column rule, which, in its turn, was taken by means of a 'grab' and slid on to the cylinder. A ring-shaped column rule was placed between each column, and the whole type cylinder, being full, had a round plate of iron placed at its end, and pressed against the type by means of nuts and screws.

Wilkinson, if his account be accurate, had thought out this complicated system as early as 1818. The new press at the *Sun* must have worked as Wilkinson produced for the Congress inquiry a copy of the newspaper, dated 25 August 1841,

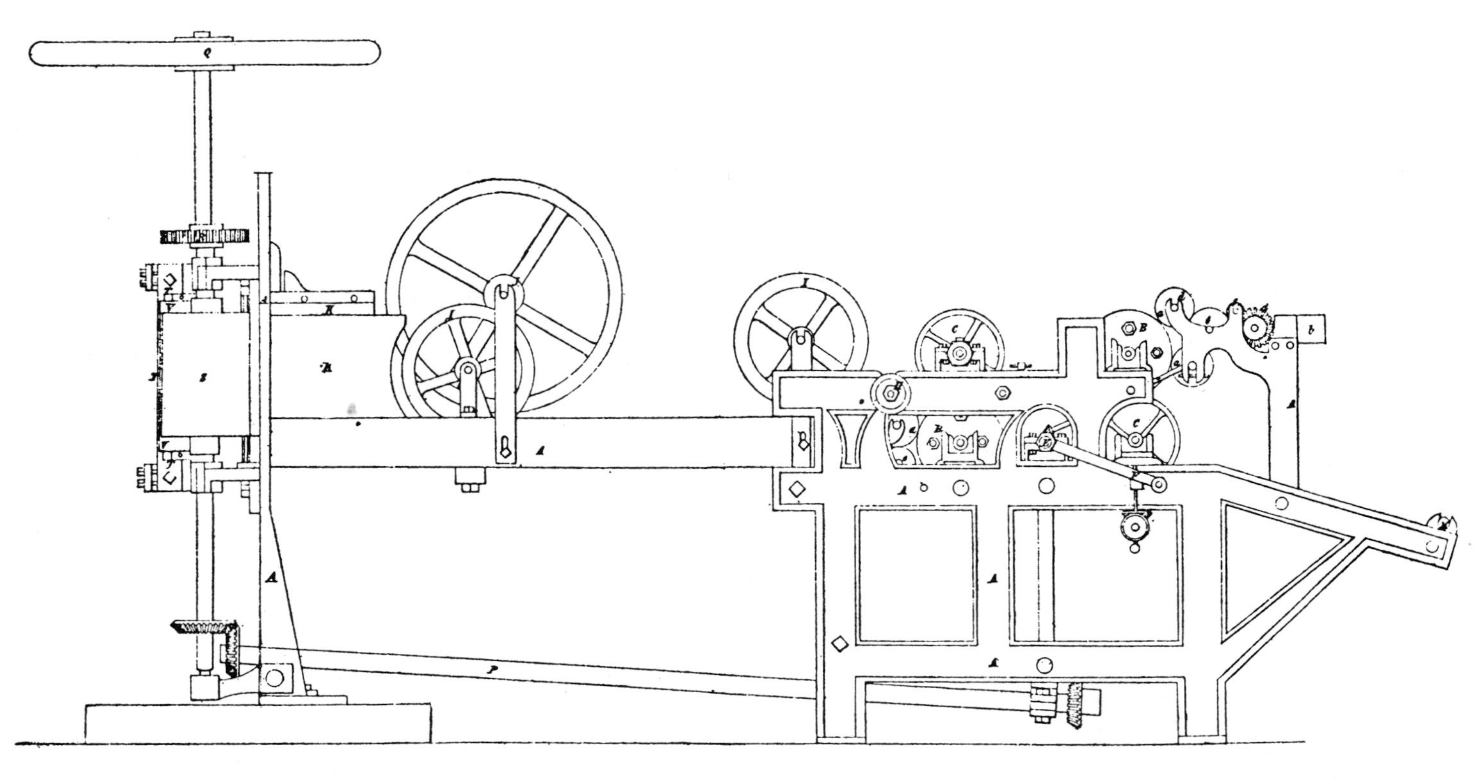

78. Diagram of rotary press patented by Moses S. Beach

noting that it was printed as a trial of his rotary cylinder press. In 1853 Wilkinson received a United States patent for a press similar to that patented in England by Beach, having in 1852 printed further copies of *The Endless Register* on an improved model press. On 7 November 1853 he printed an eight-page newspaper, again called *The Endless Register*, with a type area of $17\frac{1}{2} \times 23$ inches, on a large press which he started to build in 1852. It was stated that it produced newspapers so rapidly that the cutting apparatus could not keep up with the output. However, the products of Wilkinson's presses are not very good pieces of printing. He patented some improvements, went to Europe on an unsuccessful selling trip, and gave up after a fire destroyed his shop on his return. A number of other American inventors patented web-fed rotary presses during the decade 1850–60 but nothing came of them.

During 1850 Thomas Nelson, junior (of Nelson & Sons, Edinburgh), constructed a working model of a rotary machine for bookwork, to be shown at the Great Exhibition of 1851 (Plate XLII). It worked, and is said to have produced some 10,000 sheets an hour, printed on both sides. Printing was from a web of paper, supplied by Cowan & Co. The reason the press was taken no further was that the inventor considered it was not capable of printing sufficiently well for the purpose for which it was intended, that is bookwork, and its possibilities for newspaper work seem to have been ignored at the time. Nelson's model survives and has been exhibited on a number of occasions.

It was clear that printing from curved plates was desirable, but a flexible mould needed to be developed. In a pamphlet of 1822, *Précis sur la stéréotypie* (printed by Imprimerie Stéréotype de Cosson, Paris), A. M. de Paroy refers to his secret method, which may possibly be an early reference to the paper-mould process, but in 1829 Jean-Baptiste Genoux, of Lyons, made a 'flan' from layers of paper, with a composition of clay and plaster in between. A flan was an open custard tart sold in Paris, and the word is still used to describe the material from which curved newspaper stereotype plates are made; in English the word has become 'flong', possibly from Kronheim's use of another word, 'flanc', which will be referred to. The 'flan' was placed over a forme of type and beaten or rolled to obtain an impression, from which a duplicate plate could be cast. This was the origin of the so-called 'papier mâché' process, although, strictly speaking, the term, which dates from at least 1753, should apply to pulped paper used for moulding.

Moses Poole patented a similar material in England on 20 July 1839. His 'flexible mould' was made of layers of paper, glue, paste, potter's earth and tissue paper to a thickness of an eighth of an inch. He was not specific about the method of casting. Joseph Kronheim, a London wood engraver, in 1843, saw at the firm of Demat, publishers, printers, type and stereofounders in Brussels, a plate being made for press, and recognized that it was not being made by the traditional plaster process. He thought the discovery important enough to introduce into England and took out a patent on 29 July 1844. His 'flanc' consisted of alternate layers of tissue paper and paste on a foundation of brown paper, which, when moist, was pressed on the forme of type. The second part of his patent is more helpful than Poole's. It reads: 'The apparatus for holding the said matrices and performing the operation of casting, composed of metal plates hinged together like a portfolio. It turns on an axle beneath one of the sides (the "lower plate"). Both plates are turned horizontal, and the matrix powdered with talc, having been laid on the lower plate, and metal strips to regulate the thickness of the casting having been laid on its margin, the upper one is brought down and clamped to it. The mould is then turned vertical, and the metal poured into the mouth formed by the bevelled edges of the plates. The upper part is covered with a sheet of paper to ensure the parallelism of the stereotype plate at the back by allowing the escape of air.'

Kronheim told John Southward, the printing-trade journalist, the facts about his discovery in a letter dated 24 August 1883, when Southward was collecting information for a history of stereotyping, and added that he sold the patent to a firm which went bankrupt.

The paper or flexible mould was thus generally known after 1830, but not until some fifteen years later does there appear to have been any move towards using this 'flan' or 'flong' to make curved plates, as without a satisfactory rotary press there was little point in pursuing this technique. However, two Frenchmen, M. Worms and M. Philippe, can be considered the pioneers of rotary printing using stereotyped plates made from flongs. Worms was a printer of Argenteuil who took out a patent in 1845 with Philippe, a mechanic, for a cylindrical machine for printing with circular plates. The press was fed with an 80-metre roll of paper, the manufacture of which was an event in the paper trade of those days. Worms and Philippe had realized that Genoux's invention could be utilized for making curved plates by bending the flexible moulds and casting in a special device, which they designed. Their press was simpler than that of Wilkinson/Beach, as the paper went straight across the machine, with one type cylinder above and the other below. The patent was also taken out in England by Gerard John de Witte on 7 March 1850. This is not very helpful when it comes to the actual casting mechanism which looks, from a French diagram, as if it were a very simple affair for pouring in the molten metal.

According to the *Mechanics' Magazine* of 13 April 1850, a demonstration of the Worms and Philippe rotary press had been held the month before in Paris, and 'the stereotype cylinder was got up in exactly fifteen minutes, and the printing on both sides quite perfect, the speed was 15,000 copies per hour, which can be augmented by corresponding steam power'. Printing was on endless paper, not wetted, and each copy was cut off with mechanical precision. The journal *La Presse* (which adopted the flong method of stereotyping in 1852) was reported to have given the first order for one of the presses, which it was claimed would enable two men to do the work of fifteen.

It is probable that the reports were too enthusiastic. The fate of the Worms and Philippe press at *La Presse* if, indeed, one was installed is not known, but it could not have lasted long, for by 1855 the newspaper was still being printed by the four-feeder cylinder presses built for it by Marinoni. The 40,000 copies were dealt with by duplicating the formes by stereotyping, and printing on as many presses as required, but this would have been stereotyping in the flat and not on the curve.

A further Worms and Philippe patent followed in 1855, with an improved apparatus for making the curved plates, and, according to A.-L. Monet (*Les Machines et appareils typographiques*, Paris, 1878), a company was formed to exploit Worms's patent, but with incomplete results. Monet also mentions other contemporary inventors of web-fed machines, including a M. d'Ardenne and M. Giraudot, but he indicates why there was little progress. Apart from difficulties in obtaining good cylindrical plates, there was also the inhibition (as in Britain) caused by the French Government's tax on single sheets of paper. Moreover, at the time, about thirty years before he wrote, double-cylinder machines could quite easily cope with the edition run of most French newspapers.

This situation applied elsewhere, even in London and New York for a time, but at *The Times*, in London, it was increasingly felt that even Applegath's four-feeder cylinder machine was not productive enough to keep pace with growing circulation. Applegath therefore turned to the rotary printing principle, without, however, the benefit of curved plates, or of feeding from the reel of paper. In 1848 he devised a large 'type-revolving' machine, which for some twenty years set the pattern for large-scale newspaper presses.

79. Hoe's ten-feeder type-revolving machine printing the *Sun*

13

THE TYPE-REVOLVING MACHINE AND THE DEVELOPMENT OF THE MODERN ROTARY

APPLEGATH'S FOUR-FEEDER CYLINDER machine at *The Times* was by no means perfect. The repeated jarring—a common early complaint concerning cylinder presses—threw the impression out of adjustment, and greater speeds were impossible of attainment if a flat bed, weighing half a ton or more, had to be propelled backwards and forwards for each impression. Rotary printing seemed the answer, even if it was still necessary to print from ordinary printers' type for the time being.

The eight-feeder vertical rotary 'type-revolving' press designed for *The Times* by Augustus Applegath, took about two years to build, and started work in October 1848, printing initially at the rate of 8,000 impressions an hour. Improvements were made, and a rate of 10,000 impressions an hour was achieved, but this was always subject to the skill of the hand-feeders—a point made by Applegath in a mildly ironic letter about Hoe's press, which he wrote in 1851, and which will be quoted.

Applegath's machine consisted of a vertical cylinder, 200 inches in circumference, on which ordinary type formes were placed, the surface of the type forming a polygon, in much the same way as the earlier Bacon and Donkin machine. The central type cylinder was surrounded by eight other cylinders, each of about 13 inches in diameter, round which paper was led by tapes, each impression cylinder having a feeding apparatus and two boys in attendance. The ink rollers were also vertical, distributing the ink on to the central cylinder. Sheets of paper were fed down by hand from eight flat, horizontal feed-boards through tapes, and were then grasped by another set of tapes and passed sideways between the impression cylinder and the type cylinder, thus producing sheets printed on one side. The impression cylinder delivered them still in a vertical position into the hands of boys, one being stationed at each cylinder to receive them.

There was a slight advantage in using the flat formes of type at first, since printing was still being carried out on cylinder machines and it was thought best not to disturb the economy of the composing room by having two kinds of type composition. However, the gaps in the type surface were a nuisance and the irregularities had to be made up by pasting strips of paper under the blankets on the impression cylinders to take up the inequalities.

Applegath was not unaware of the drawback of printing from the polygonal surface or the inconvenience caused to feeding and delivery by the large type cylinder, and so he went to work on a machine with two smaller type cylinders, only 70 inches in circumference, which could utilize either ordinary printers' type, or tapered type which could be locked around the cylinders. He also produced an eight-feeder for *The Standard* and another, nine-feeder, for *The Times*. The actual building of the machines was carried out by Thomas Middleton, engineer, of Southwark, who made a small, four-feeder on Applegath lines for the *Illustrated London News*, which was shown at the Great Exhibition of 1851. Middleton claimed that the eight-feeder averaged 12,000 impressions an hour and the nine-feeder 16,000 impressions an hour. Middleton also made a ten-feeder vertical machine 'originally projected by Applegath' for the *Morning Herald*, which was adapted to work with stereotype plates and was said to have had an output of 20,000 impressions an hour.

Before this happened, however, Applegath's machine had been severely challenged in New York. While Applegath had been working on his vertical cylinder rotary, the firm of Hoe, during 1845 and 1846, concentrated on an invention of Richard Hoe based on a horizontal type cylinder. The 'Hoe Type Revolving Machine' could be made in sizes from two to ten cylinders. The type cylinder was about 6½ feet in diameter and about a sixth of its circumference constituted the printing surface, the remainder acting as a cylindrical ink-distributing table, which it is reasonable to assume Hoe got from Napier's invention.

Each page was locked up on a detached segment of the cylinder called by the compositors a 'turtle'. The type was ordinary printers' type, but wedge-shaped column rules bound the type near the top and were held down on the turtle by tongues projecting at intervals along their length, which slid into rebated grooves cut crosswise in the base of the turtle. The space in the grooves between the column rules was filled with sliding blocks, and screws at the end and side of each 'page' locked the whole together. It was claimed that the types were as secure on the cylinder as on the old flat bed, but if locking up were imperfect when the machine was put in motion centrifugal force would spray the loose type around the machine to the danger of the operatives.

Sheets were fed in by boys and taken from the feed boards by automatic grippers operated by cams in the impression cylinders, and which conveyed them around the revolving forme on the central cylinder. After printing, the sheet was conducted underneath each feed-board by means of tapes to sheet flyers which laid them in piles on tables. This was an advance on the Applegath machine where a taker-off boy was required in addition to each layer-on.

Running speed was claimed at 2,000 sheets per feeder an hour—thus a 'four-feeder' would have a capacity of about 8,000 sheets printed on one side an hour. Applegath complained (in a letter dated 8 March 1851) about a report in *The Expositor*, which had stated that a Hoe eight-feeder would give 20,000 impressions an hour, or 2,500 from each feeding position. He wrote: 'Why the New York layers-on can place 2500 sheets *larger than the Times*, while our hands have only attained to 1500, I cannot pretend to determine; but *the rate of motion* being slower in the *Times*' machine than in Mr. Hoe's (at the same number of revolutions), the

80. Applegath's eight-feeder vertical type-revolving machine. (See page 185)

superior production of the New York machine must be explained on other principles than an inferiority of mechanical means.

'The machines at the *Times* are one variety only of the vertical system; they are made to print *flat columns of type*, in order not to disturb the economy of the office by having two sorts of composition; but in the specification of my patent, sheet C, another form of machine is given purely cylindric, yet using ordinary type, which, by the addition of an upper story of feeders, is capable of printing both forms of the *Times* at the rate of 22,500 sheets per hour, or 45,000 impressions.

'This machine I had the honour of offering to the authorities of the Great Exhibition, stating the produce at 40,000; but, could I depend upon obtaining layers-on with the superior sleight of hand of the New York men, the estimate might be increased—until which period I must be contented with data obtained by my own actual experience.'

The first of Hoe's type-revolving machines was installed at the Philadelphia *Public Ledger* and, in 1848, *La Patrie*, in Paris, ordered a four-feeder. This was an unfortunate year as the disturbances arising from the fall of the monarchy and the establishment of the 2nd Republic, and the reimposition of a newspaper stamp duty, inhibited further newspaper publishing enterprise. But the machine was kept working, and on a visit to Paris, Edward Lloyd of *Lloyd's Weekly Newspaper*, of

81. A side view of the Hoe ten-feeder type-revolving machine

London, saw the machine in the office of *La Patrie*. He ordered a six-feeder version for his office in Salisbury Square, Fleet Street, the installation taking place in 1856.

The French political upheavals of 1848 had another side-effect for newspapers. The brothers Dellagana had been producing 'flongs' for the *Constitutionelle* newspaper in Paris. Whether they were driven out of the country, as is sometimes stated, or they simply felt there was no future for them in the French press, they left for London, where they set up a stereotyping foundry using their knowledge of the 'papier mâché' process.

The proprietor of *The Times*, John Walter III, had not overlooked the possibility of making curved stereotyped plates. His father, John Walter II, had tried to make moulds of composed pages with the assistance of Marc Isambard Brunel in 1819, Brunel taking out a patent the next year. He used a layer of composition made of pipeclay, chalk and starch which was spread on a perforated steel plate hinged to the galley of type and brought repeatedly down on to the type to make a mould. The patent envisaged curved plates even at that early date, as the following extract indicates: 'The mould when dry, retains sufficient flexibility to assume a cylindrical form, and plates of that shape may be cast by placing the mould in a box formed of parallel plates of the required radius.' However advanced in thought, the process did not work out in practice, and *The Times* management—John Walter III and his chief engineer, John Cameron Macdonald (appointed in 1856)—had to wait until the 'papier mâché' process was introduced to England and for an experienced man to help them.

Whether Poole or Kronheim should be given the credit for introducing the 'papier mâché' method to England can be debated, but a great number of experiments were needed before it could be used to make satisfactory curved plates, and *The*

Times called upon James Dellagana to assist them. Experiments began by casting columns of type and arranging them, when planed and finished, in a forme of four pages, which was worked off on the cylinder flatbed machine, producing nearly 5,000 extra impressions an hour. This was encouraging, and the next step was to adapt such stereotyped columns to the polygonal chases on Applegath's vertical rotary press. This achieved another 5,000 impressions an hour, and the stage was set for the final effort of taking a 'papier mâché' matrix from a whole page at one operation. A casting box was devised, curved to the circumference of the printing cylinder, but although printing from a curved stereotyped plate took place on 28 December 1857, experiments continued until 1859 before this became a regular practice, and it was not until March 1860 that Dellagana was instructed to make a stereotyping department a regular constituent of *The Times* establishment.

Applegath had been agitating to go horizontal, and Walter instructed Macdonald to get a new machine in the early part of 1857—but it was to be a Hoe. Walter had been shocked when he learned that Lloyd had ordered a Hoe six-feeder machine, and had journeyed to Philadelphia to visit the machine room of the *Public Ledger*. As a result, he favoured the Hoe machine and ordered two ten-feeders at £6,000 each, which, later, Macdonald got reduced to £10,000 for the two. The machines were installed in August 1858, and Applegath's verticals were broken up.

The Hoe machines for *The Times* were made in England at the Manchester works of Sir Joseph Whitworth & Co., and Hoe became aware of the importance of the British market, eventually setting up a branch in London in 1865. Orders for Hoe

82. The Hoe six-feeder type-revolving machine

type-revolving machines were received from leading newspapers and at such offices as the *Daily Telegraph* and the London *Standard* as many as five were in operation.

Hoe also made a type-revolving book perfecting machine in four sizes, from 24 × 27 inches to 33 × 50 inches, which would print from 1,500 to 2,000 perfected sheets an hour. This was an indication that a perfecting type-revolving press for newspapers was not an impossibility, but by the time thoughts were turning to printing news sheets on both sides simultaneously the presses had taken a different form—the web-fed rotary, which in Britain became a commercial possibility after the repeal of the paper tax in 1860.

Type-revolving presses continued to be made in France until 1871 when the newspaper stamp was abolished. The principal manufacturer was Marinoni, of Paris, who set himself the task in 1877 of improving on Hoe's type-revolving machine. He constructed a six-feeder for the *Petit Journal*, which differed from other newspaper machines of its kind in that it perfected the sheet; although, as noted, Hoe made a book machine on these lines. Each layer-on fed 1,500 sheets an hour, so that for six boards the product was 9,000 sheets, printed on both sides. The sheets were taken off by flyers, of which there were four sets. The machine was successful, being sold outside France. One was installed at the London *Echo*, and the management was so satisfied that when Marinoni brought out his web-fed rotary, the *Echo* was among the first to make an installation.

But the multiplication of the number of impression cylinders to increase the rate of printing could not go on indefinitely. It was found that with the multi-feeder machines difficulties arose, and that stoppages were frequent, for human beings are not automatons. The solution to the feeding problem was obviously to do away with 'layer-on' and feed from an endless roll of paper. When, in 1860, the British paper tax was repealed, and the problem of making curved stereos appeared to be on its way to being solved, John Walter III began thinking about a reel-fed, perfecting rotary press, and early in 1862 he authorized the commencement of experiments, in which the most arduous task, as it turned out, was the establishment of precision in the making of the curved stereotype plates, at which Dellagana worked for three years up to 1866. Because flat stereotype plates were made from 'flong' and curved plates had been cast for the large cylinder of the Hoe machine, it did not follow that the semicircular plates required for the new rotary could be produced with accuracy, in quantity and at speed. However, difficulties were overcome, and by 1868 Walter was in a position to order three finished machines, which were installed at the end of 1869.

While the work was going on at *The Times* in London, an American, William Bullock, of Philadelphia, had patented a rotary press in 1863, constructing the first model for the *Philadelphia Inquirer* in 1865. This may well be claimed as the first automatic, reel-fed rotary press working from stereotype plates, and printing on both sides of the paper. The machine consisted of two printing and two impression cylinders, the second of these being very large to lessen the set-off from the first printed side of the paper. The paper, fed from the reel, was cut into sheets before it reached the impression, the sheets then being carried through the press by tapes and fingers. The press was capable of delivering about 10,000 flat sheets, printed on both sides, an hour, but difficulty was experienced in maintaining this

83. William Bullock's reel-fed rotary machine, 1865

speed at the delivery end, as the simple 'fly' could not handle more than 8,000 sheets an hour.

In some respects, for the smaller newspaper, for example, the cutting of the paper before printing was an advantage, since common paper could be used and, there being no tension, the breaking of the web in its course through the machine avoided. The avoidance of set-off meant that special ink was not required either. But, in the long run the productive capacity required by large-scale newspapers was bound to lead to demands for printing on the web, for strengthened paper and quick-drying rotary inks.

Bullock did not live to see improvements made to his press. He was involved in a fatal accident in the gearing, and died on 12 April 1867. Speed of delivery was improved by the automatic folder, invented by Walter Scott, of New York, in 1869, and with a device, patented by John William Kellberg, which supplied the paper to the cylinders without tapes, cords or grippers and situated the knife after the printing cylinders. With the addition of this equipment, it was claimed that 20,000 perfected sheets an hour could be produced. The *Daily Telegraph* was among the newspapers, in 1870, which installed the improved version of the Bullock press. In time, however, the Bullock rotary was superseded by the *Walter* and the *Hoe*.

The *Walter* machine, as that at *The Times* was called, set the pattern for subsequent rotary presses. The roll of paper was placed at the end of the machine on two standards and the web passed over a tension roller and then over and under two damping cylinders. To prevent any drag in the printing, the paper was then conducted between two small rollers, the surfaces of which travelled at the same speed as the impression cylinders. The printing and impression cylinders were arranged one above the other, the top and bottom ones carrying the plates, those in the centre being used for the impression. The paper passed over and under the centre cylinders; the side of the paper being thus reversed in the operation. In order that no perceptible set-off should occur a large surface-drum of the same diameter as the second impression cylinder worked in contact with the latter, removing the ink which accumulated on the blanket.

After the paper was printed on both sides it passed directly to the cutting cylinder, where the sheets were not entirely separated, being still joined by two narrow strips. The sheets were then directed up a series of tapes to reach two small rollers which tore a sheet from that following. The sheets were then conducted between two rollers and by a dividing motion were directed perpendicularly down a series of tapes, of which there were two sets, to receive the paper alternately when the flyers struck them down to the board on either side. At this point, there was no folding apparatus and for some years newspapers were delivered unfolded. The average speed claimed for the *Walter* was 12,000 perfect copies an hour. This machine initiated modern newspaper printing and served *The Times* until 1895. The *Walter* press was not restricted to *The Times*, as it was used by *The Scotsman*, the London *Daily News*, the *Birmingham Post*, and even by the New York *Times*, in the heart of the Hoe territory, as it were.

84. The Walter rotary machine

Hoe's took until 1873, in fact, to produce a web-fed rotary press, but, by that time, were able to take advantage of a number of developments which had taken place to turn out the forerunner of perhaps the best-known nineteenth-century newspaper press. Hoe did not mind admitting that he got ideas from other inventors, and was quite willing to buy up patents when necessary. His firm also had the manufacturing capacity denied to *The Times*, which, after all, was primarily in the business of publishing and not machinery manufacture.

Before, however, the coming of the few giants in rotary-press manufacture, a number of small engineering firms, and even publishing houses, ventured into the field, and, in the process, often introduced innovations which were taken up by others. There were also individual inventions which contributed to the final form of the rotary press. Thomas Jefferson Mayall's 'accumulating cylinder' of 1867, and Hedderwick's apparatus of 1870 for counting the printed sheets into dozens, quires or other required numbers, are two examples. Equipment for forming stereotype moulds were numerous after 1867, and in one patent—that of Dmitri Timiriazeff—there appears the first mention of 'papier mâché' in so many words.

Percy David Hedderwick was the son of the proprietor of the Glasgow *Citizen* and an ingenious inventor, who patented an advanced rotary press in 1870. It proved too complicated, and the *Citizen*, which from 1867 had been printed on a six-feeder Marinoni cylinder machine, went on to use Marinoni's web-fed rotary when he put it on the market.

Among the first of the small, successful rotary-press builders were Duncan and Wilson, of Liverpool, who in 1870 produced their first *Victory* machine for the Glasgow *Star*, and associated newspapers, and subsequently for English provincial newspapers and the London *Globe*. The credit for bringing a folding apparatus into one unit with a printing press belongs to George Wilson, and this was one of the outstanding characteristics of the *Victory* machine. From the printing cylinder the reel of paper proceeded to a folding blade, at which point the paper was cut across, the blade descending and pushing the sheet to folding rollers, whence it was conveyed between tapes to a delivery board in a pit below the press. The manufacturers were aware that smaller newspaper houses could not afford a stereotyping foundry, and accordingly the *Victory* was adjusted so that printing could be from either movable type or from stereotype plates.

The *Victory* was popular, and by 1879 more than fifty had been installed. Capacity was about 10,000 copies an hour. The machine also earned itself a small niche in printing history as it was used in an early experiment in linking of machines. In New York one machine was responsible for the printing and folding of a newspaper, the back of which was pasted. A duplicate machine, driven by gearing from the larger one, printed a wrapper, which was conducted by means of tapes to the folded sheet and a pasted-up newpaper was delivered printed, folded and bound in a neat cover by the dual machine.

The *Prestonian* rotary machine soon followed. This was the invention of Walter Bond, foreman of the *Preston Guardian*, aided by a Preston engineer, Joseph Foster, who developed into one of the bigger printing-press manufacturers. Like the *Victory*, the *Prestonian* could be used for printing from either type or stereotyped plates, or, surprisingly, from both combined, a facility which was considered to be a 'decided advantage' to a newspaper, since items of news could be frequently inserted. This technique can be seen as a forerunner of the 'stop press' and 'fudge' system of dealing with late news. The type cylinders were 3 feet 6 inches in diameter—the large size enabling the movable type to be used. The removable curved type 'formes' occupied a quarter of the surface of each cylinder, the rest acting as an additional inking plate. Adjacent to each type cylinder were four impression cylinders. The web of paper was first printed on the inner cylinder, and, as the forme occupied only a quarter of its circumference at each revolution four successive impressions were made. The paper then passed to the outer cylinder and was perfected in a similar manner. The printed web passed between cutting cylinders and was perforated across its entire width, and finally severed at the point of perforation. Sheets were deposited by a flyer on to a receiving-board.

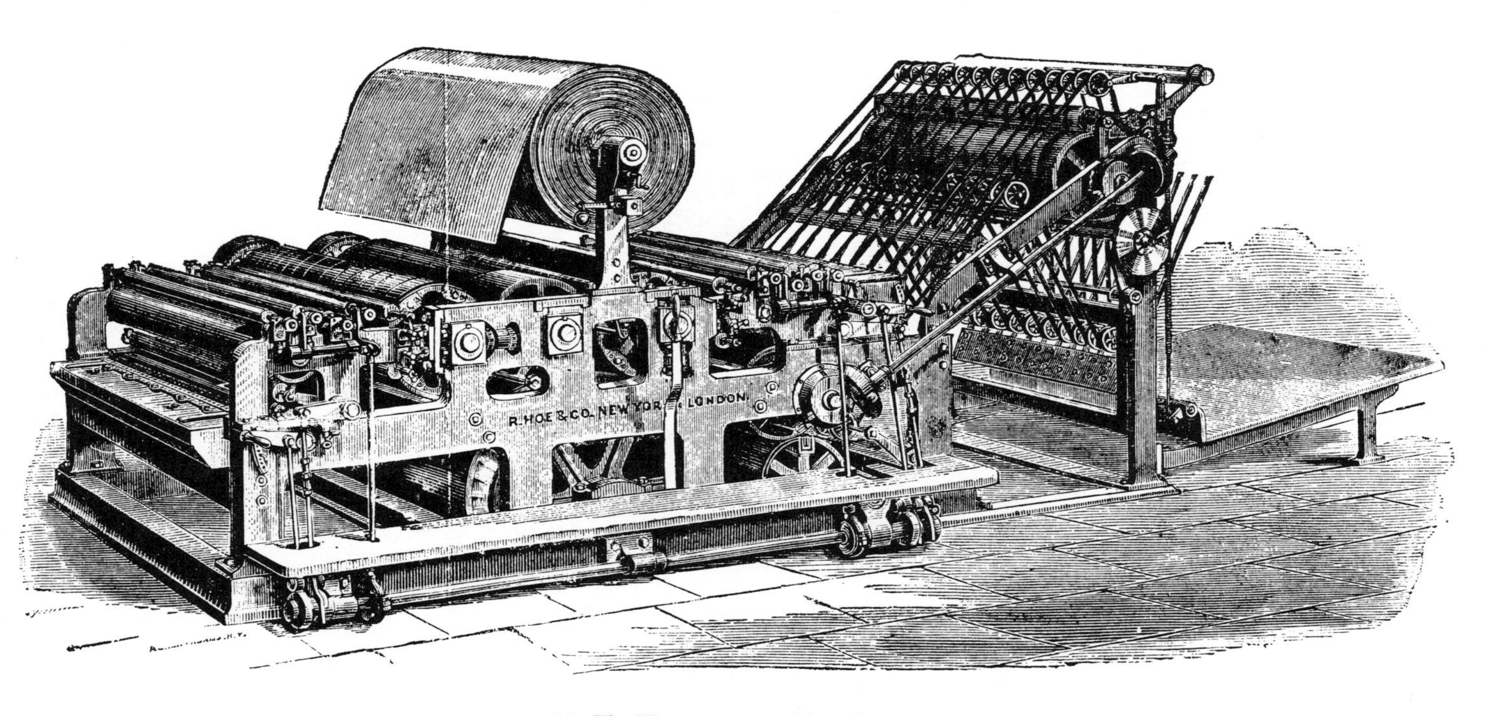

85. The Hoe rotary machine of 1873

Later, a folding machine was added to the *Prestonian*. Speed was between 12,000 and 14,000 perfected sheets an hour, or from 10,000 to 12,000 copies an hour, folded. The machine was most popular in the English provinces, but two London newspapers, the *Standard* and the *Evening News*, also installed *Prestonians*.

Another English provincial rotary was the *Northumbrian*, made by Francis Donnison & Son of Newcastle upon Tyne, and used mostly in the north of England and in Scotland. Among newspapers which installed the *Northumbrian* were the *Greenock Telegraph*, the *Edinburgh Evening News*, the *Sunderland Daily Post* and the *Dundee Evening News*. The machine consisted of six cylinders arranged parallel to each other in a horizontal frame, those at each end being used for ink distribution. Next to the ink-drums were the type-cylinders, the impression being given by the two in the centre. The paper was placed at either end of the machine above the cylinders in standards, and when one reel was exhausted the other was quickly attached. The paper passed horizontally from the reel over a roller directly above the centre of the machine, then on to the first impression cylinder and from thence to the second. It was then conducted to the cutting cylinders below, when, by means of an oscillating frame or divider, it was directed alternately into sets of delivery tapes on either side. The sheets were then struck down to the taking-off boards by means of flyers. One improvement in this machine over others concerned the unwinding of the reel, which was assisted by a small roller placed directly above and working upon the inner forme cylinder. By this the drag was considerably reduced, the small roller performing the work which was mostly done by the cylinders as the impression was taken.

A machine large enough to print a daily newspaper measured 14 × 5 feet, and was about 5 feet 6 inches high, and as there was no gearing below the base of the frame no pit was required. This was therefore a highly convenient small machine, but its speed was limited to 9,000 perfected copies an hour—a great advance on that of the earlier presses, but in the light of what was to develop not very great. However, Foster persevered, and a more massive successor to the *Prestonian* (20 feet long, 7½ feet wide and 7½ feet high), for which a pit was found desirable for maintenance purposes, was produced. This could deliver 12,000 eight-page papers an hour flat, or slightly less if a folder was attached. A machine for use with stereotype plates was known simply as the *Foster*, and a rotary bill machine for printing newspaper contents bills could print from special type cut on the curve so that it made the required diameter of the cylinder when in place. It ran at the rate of 14,000 copies an hour.

The enterprising firm of Hoe was not going to be left behind in the rotary-press race, and in 1873 made its first web-fed perfecting rotary machine, which, though in appearance similar to others, differed in a number of material points. It was somewhat bigger, being about 19 feet long by 8 feet wide and 8 feet high, and the roll of paper was placed immediately above the type cylinders, which were fitted in a horizontal frame. The second impression cylinder was three times the diameter of the first, and, unlike the latter, which was parallel with the type cylinder, was located underneath. The advantage of having the large type cylinder was that the impression was given on three different portions, thus obviating, to some extent, the set-off. The sheet, after having been printed by the inner forme, passed under the cylinder to the outer, and thence between two cutting cylinders, which were the same in diameter as the type cylinders. The knife did not completely sever the sheets, a process which was effected by a separate series of tapes which travelled faster than those on the main portion of the machine, and dragged the sheet away from the succeeding one, placing a regulated space between them.

The sheets travelled over a drum with a slightly greater circumference than the length of the paper and when any desired number of sheets were gathered they were directed by a switch down the flyers, and deposited on the taking-off board. This gathering and delivering cylinder was patented by Stephen D. Tucker, of Hoe's staff, and solved the problem of rapid flat delivery. He went on, however, to look into the question of the delivery of folded newspapers.

The first newspaper proprietor to install a Hoe web perfecting rotary (made in the London factory) was Edward Lloyd, who had been the first to challenge *The Times* with a Hoe type-revolving machine. At *Lloyd's* a plan for dividing the sheets into quires was adopted—a suggestion of later developments. On the delivery of every twenty-seven sheets (which was the regulation quire at *Lloyd's*) the table was slightly moved by means of cams to the right or left, and the quires could be easily and quickly separated.

The inking apparatus consisted of two drums parallel to each other, and they were each provided with a series of vibrators and wavers. These were placed in the horizontal side-frame on the same level as the type cylinders. The machine printed on average 14,400 perfected sheets an hour flat, but folding devices could also be provided, as in 1875 Stephen Tucker had patented a rotating folding cylinder which folded papers as fast as they came from the press, or about 15,000 an hour. These folders were also used on the *Bullock* press as were those made by C. Potter Jr. & Co.

They were, however, only the first in a series of experiments, consisting of the combination of a 'gathering cylinder' with a rotary folding cylinder and tapes to convey the printed sheets under horizontal folding blades, which thrust them at the appropriate point between folding rollers placed at alternate angles, before delivering them on travelling belts by a small flyer. In 1875 Edwyn Anthony and William Wilberforce Taylor, in England, took out patents for 'turning surfaces', by which the webs of paper could be turned and twisted in any direction required for printing. Hoe, who was in England at the time, saw the possibility of those patents and purchased them for both Britain and the United States.

In their various experiments the firm of Hoe found that they had encroached on patents secured by Luther C. Crowell, of Boston, who had made a machine for making paper bags. His folding-machine patents were secured by purchase in 1877, and one, in particular, enabled Hoe to expand the machines to accommodate two pages across the cylinder. This was the popularly named 'cow catcher', patented by Crowell in 1883, being a triangular piece, now known as a 'former', which folded the web from its full width to half width in the direction of its length.

The Hoe rotary was first introduced in the United States at the New York *Tribune*, followed by the Philadelphia *Times*. In Britain *Lloyd's* was followed by the *Standard*, the *Daily Telegraph* and the *Liverpool Mercury*.

Other manufacturers entered the rotary-press field in Britain, Europe and

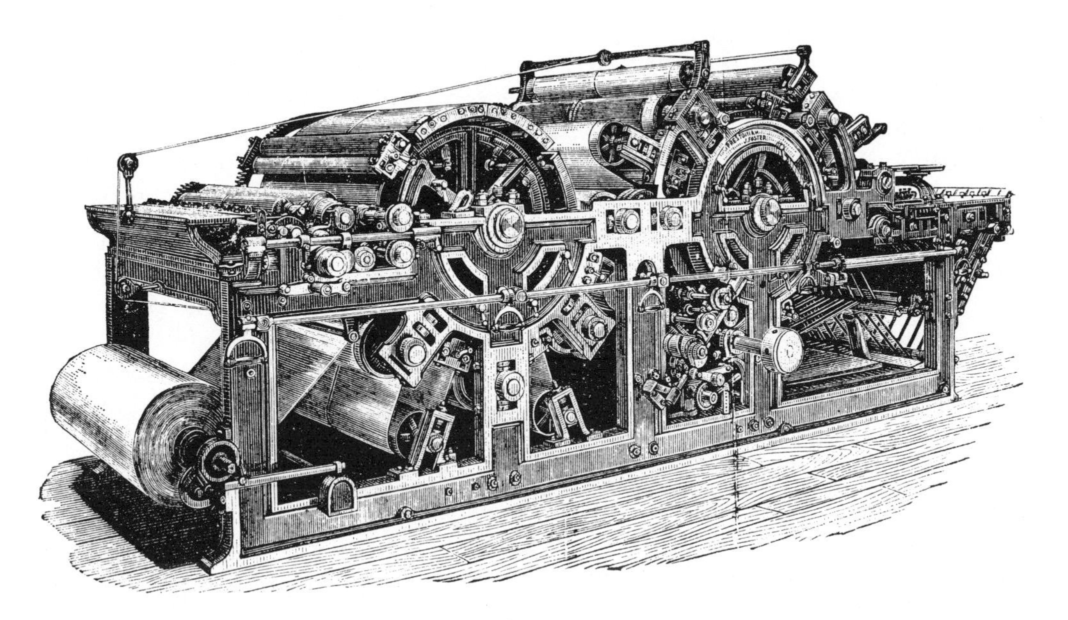

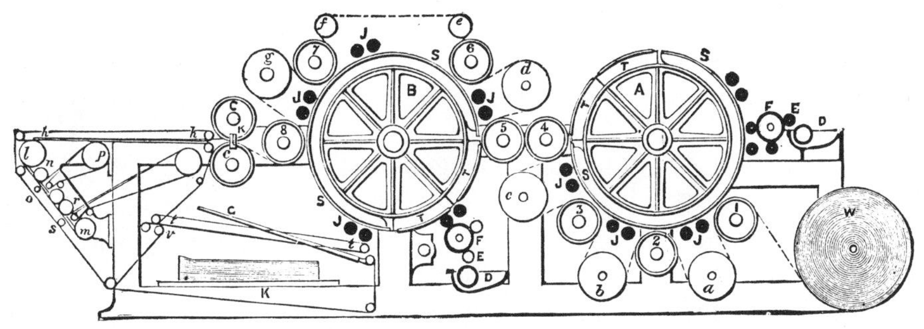

86. The Prestonian rotary machine with (below) a sectional view. (See page 193)

America, sometimes, as in the case of the *Whitefriars* and *Ingram* machines, being inspired by a particular printing or periodical office. During the 1870s the newspaper and periodical side of the printing industry was expanding rapidly as public demand for reading matter grew, but the printers involved were not yet habituated to the idea of ordering, as a matter of course, stock machines from a manufacturer. They still wanted machines made for them. Joseph Pardoe, superintendent of the Whitefriars printing works of Bradbury, Agnew & Co., decided, in 1873, to produce his own rotary with the aid of B. W. Davis, the engineer of Lower Kennington Lane in south-east London. Named the *Whitefriars* after the printing works the first machine was erected to print the *Weekly Budget.*

At first the machine was meant to be a fast perfecting two-feeder, and laying-on boards were provided directly above the cylinders. In this way the *Whitefriars* can be considered an early example of the 'sheet-fed rotary' which became popular nearly a hundred years later. However, Pardoe and Davis decided to incorporate a reel-feeding device as well. With hand-feeding a speed of 4,000 impressions an hour was achieved, the figure being doubled when the reel was used. Of light construction, the *Whitefriars* consisted of four cylinders arranged around a semicircular frame, which gave it a look quite different to the other rotaries which had been developed. It was described as 'horseshoe-shaped'. The two centre drums were used for the impression, while the outside cylinders received the plates, which could be either cast or bent in a curvilinear form. This was also an unusual departure. The cylinders intended to receive the plates were curved spirally, with grooves cut and undercut to enable the screw catches to travel to any part of the cylinder and hold the plate securely—the size of the plate making no difference. Unless the new casting box was used, plates were cast flat and curved afterwards.

A paper matrix was made from a forme in the usual way, and when the plate had been cast and trimmed it was laid on the hot chamber near the metal pot and allowed to lie until it was soft. It was then placed in a bending box, where it received the appropriate curvature to enable it to lie on the plate cylinder. The bending box was made of iron, heated inside by several jets of gas, and was the size and shape of the printing cylinder. After curving, the plate was taken away and laid on a cooling saddle to set firm. The saddle was a section, rather like the 'turtle', and allowed the plate to cool in the correct shape.

The success of the machine may be judged by the fact that it was not restricted to the Whitefriars printing works but was also sold to about ten other printing houses. In time, however, it was superseded by the products of the larger manufacturers.

Three years after the launching of the *Whitefriars*, W. J. Ingram, proprietor of the *Illustrated London News*, felt that a special rotary was needed for printing the engravings which were so important a part of his publications. Together with his works manager, James Brister, he devised the *Ingram* rotary machine, which was built for him by Thomas Middleton, of Southwark, who had earlier built both cylinder and type-revolving machines for Ingram.

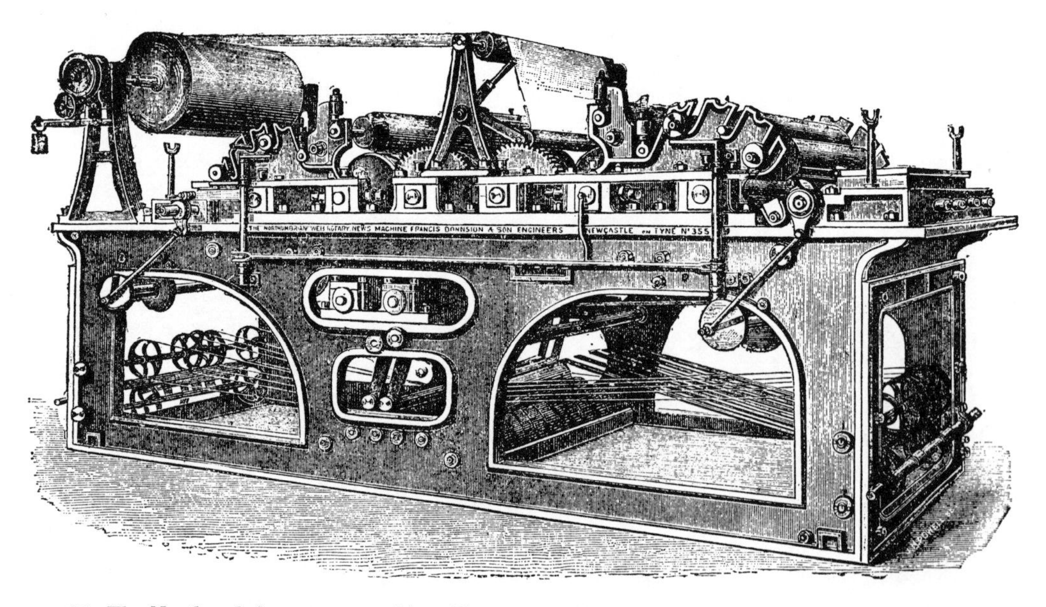

87. The Northumbrian rotary machine. (See page 195)

Ingram had found it difficult to obtain satisfactory impressions from engravings bent to the 'sharp curve required to correspond to printing cylinders of the ordinary size' and he, therefore, considerably increased the diameter of the printing cylinder to which the engraved plates were to be adapted so that the curves to which the engravings were bent might be gentler and of longer radius than the curved surface of the other printing cylinder. By this means he was also able to place on the same printing cylinder two, three or more copies of the engravings, so that while the surface speed of the large and small printing cylinders was the same, the small cylinder, if it contained only one set of stereotype plates for the letterpress, would rotate two, three or more times for every revolution of the large cylinder. The impression cylinder, which acted in conjunction with the large printing cylinder, was also correspondingly increased in size and rotated at the same surface speed. A special folding machine, made by Harrild, was attached, and the output was 7,000 copies an hour.

A firm which made a name for itself before 1890 was James Farmer, of Salford, near Manchester, and in view of the origins of rotary printing which have been outlined, it is interesting to note that Farmer was a calico printer rather than an engineer. Farmer's rotary machine was 32 feet in length and its peculiarity was its adaptation to print from two reels placed one above another, so that although no extra plates were required the production could be increased. The reels were placed on an independent frame at the end of the machine, from which the paper ran into the top of the machine, when it was conducted by a series of tape bars to the inner forme cylinder, and from thence to the outer. After the last side was printed the paper passed to the cutting and gathering frame. A special arrangement which could be regulated at will allowed the sheets, after being cut, to be collected in parcels of a dozen, when the whole received a fold in the centre and ran into a single trough. The machine was estimated to produce 20,000 copies an hour. By 1890 very few of the machines were being made.

It was to be expected that existing printing-machine manufacturers would try their hand at rotary-press building, and Dawson's, of Otley, brought out a *Wharfedale* rotary, the first being erected for the *Bradford Observer*. This was very compactly arranged. The first pair of cylinders was above the second set, but the travel of the sheet was much the same as in other machines. After printing, the sheet was perforated by a serrated knife, and passed into a folding machine which delivered two copies at a time. The folding machine, which was of smaller construction than others in use at the time, was patented by Dawson. The machine was capable of printing and folding about 9,000 copies an hour.

Rotary presses were now made in various parts of the world. Marinoni, of Paris, who had had some success with his type-revolving machines, turned to web-fed rotaries as soon as the time was ripe in 1872, and built a machine for the newspaper, *La Liberté*. It was of unusual construction, as the four cylinders were arranged immediately above each other, the two in the centre being used for the impression and the top and bottom ones for the plates. The ink ductors were situated near the plate cylinders, which were each provided with sets of rollers, the wavers having a side motion imparted by eccentric cams. After printing, the sheets were collected together on two 'gathering drums' and when five sheets were thus gathered a set

of tapes on a vibrating frame moved out and directed them down on the flyers, which deposited them on a taking-off board. The average speed was 10,000 perfect copies an hour.

An improvement was made when the London manager of the firm, A. Sauvée, patented in 1880 an arrangement by which a forme of any size could be printed, the sheets being cut before entering the printing apparatus by a knife adjusted to suit the size required.

This was one of the distinguishing features of the Marinoni, as, apart from the early Bullock, it was the only one in which the sheets were cut before instead of after printing. Kellberg's apparatus altered the situation on the Bullock. The Marinoni had four delivery boards and required no attention other than the removal of the sheets from time to time as they accumulated. Later, an 'accumulator' was added, by which the sheets were collected in quires. When a Marinoni was erected in the London office of the *Globe* it was fitted with a folding machine and this necessitated an alteration in the manner of taking-off. The gathering drums were dispensed with and the sheets, after having been cut, passed horizontally between tapes under two folding knives, which struck them down between a set of rollers, after which the last fold was administered by another having a side motion.

Another Paris manufacturer, Jules Derriey, is said to have tried to beat the newspaper-stamp laws. He took out a patent for a small, compact rotary in 1866, and built one in 1868 for *Le Petit Moniteur*, which was printed from the web. But the stamp law was invoked, and the machine was converted for hand-feeding and sold to *El Imparcial*, of Madrid. Derriey constructed a new model in 1872, which was installed at *Le Petit Moniteur*. Dimensions were only 11 feet long by 7 feet wide, and a speed of 12,000 impressions an hour was claimed. Eventually, a folding machine was attached, delivering the paper in quires, folded or flat, and in four-page or eight-page editions. A Derriey rotary press was erected in London for *Reynold's Newspaper* in 1878.

The first rotary press made in Belgium was by H. Jullien, of Brussels. This was small, and intended for the printing of periodicals.

Friedrich Koenig, inventor of the cylinder press, had envisaged rotary motion, but techniques had not sufficiently developed in his day for this to be achieved. It was left to his son, another Friedrich, to initiate the first web-fed rotary press at the Obserzell factory of Koenig and Bauer. This was built in 1875, and was supplied to the *Marburger Zeitung*. It lacked certain refinements, and was of the open-sheet-delivery type. Later, folders, as developed in the United States, were added.

In the United States, a number of rivals to Hoe had entered the business of press manufacture, which had been substantially in the hands of Hoe and the A. B. Taylor Company until about 1860, when Andrew Campbell, a former foreman at Taylor's, set up on his own, erecting a factory at Brooklyn in 1866. Campbell built a web-perfecting press for the *Cleveland Leader* as early as 1876, with a capacity of 12,000 newspapers an hour, folded and delivered automatically. Campbell left the firm, which made no further presses after 1905.

The Goss Printing Press Company produced its *Clipper* in 1885 with a capacity of 8,000 four- or eight-page newspapers an hour and in the same year its *Monitor*,

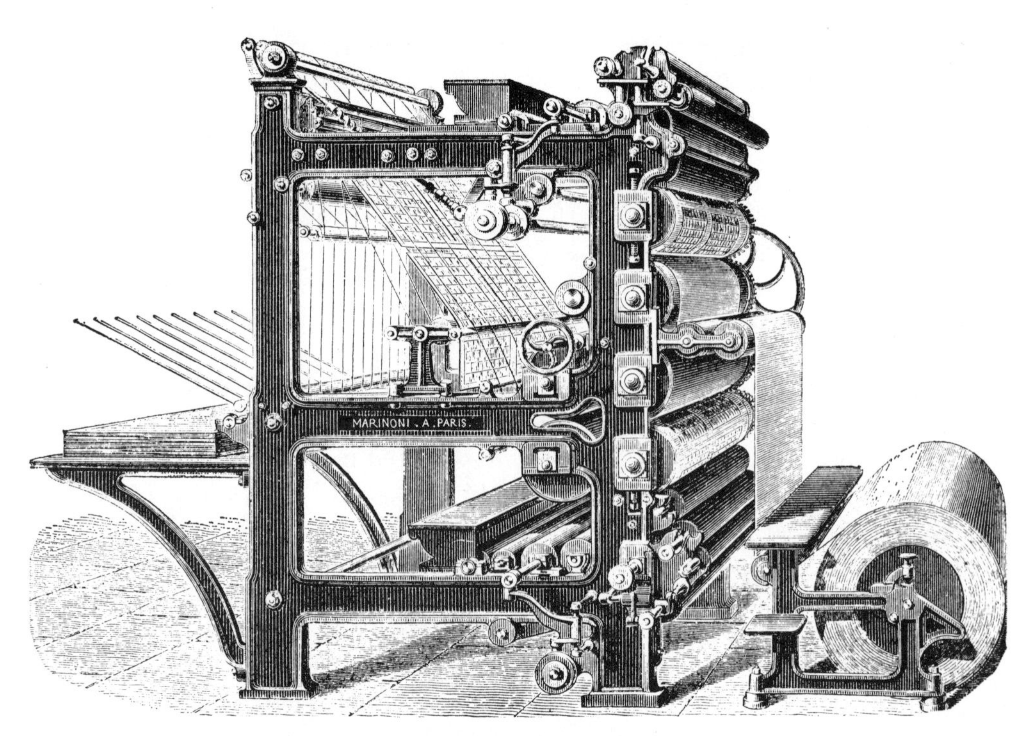

89. The Marinoni rotary machine. (See page 199)

which could produce 18,000 similar newspapers in the same time. Walter Scott & Co., of Plainfield, New Jersey, the Potter Printing Press Company and Duplex are among the names of other companies in the field.

But there was to be an interesting development, primarily in the United States and then in Britain, when it was found that some smaller newspapers did not, after all, require large rotary presses capable of printing from cast stereotype plates, although they were aware of the advantages of feeding from the reel of paper. Thus were developed the reel-fed flatbed newspaper presses, which will be dealt with before the chapter on the emergence of the multi-unit rotary presses, and the many refinements which have gone into modern rotary-press installation.

Although it is not possible to give a precise date for the beginnings of the modern conception of rotary printing, since, as with most developments, there were precursors, the decade 1890 to 1900 was important. The effect of the Education Acts was beginning to show in increased demand for popular reading matter, and new developments in half-tone reproduction, ink and paper-making all tended to produce a new situation in which the rotary press—which had been revolutionary in itself—would have to be made still more productive.

Frederick Wilson and Douglas Grey published their book *Modern Printing Machinery* in 1888, and this date is roughly the point at which new thinking began to take place about rotary installations. It is of interest therefore to reproduce their list of London newspapers and the presses employed. London (and perhaps New York) was the most advanced printing centre at the time, and what was going on there would give a fair picture of what stage rotary printing was in. The combined circulation of the daily newspapers mentioned was about 1,200,000 copies.

The Times	Eight Walter machines
Daily Telegraph	Ten Hoe machines
Standard	Six Hoe machines and six Prestonians
Daily Chronicle	Four Hoe machines
Daily News	Eight Walter machines, with cutting and folding attachments added by Foster of Preston
The Globe	Three Victory machines
Echo	Six Marinoni machines
St. James's Gazette	One Ingram and one Hoe machine
Pall Mall Gazette	Four Marinoni machines
Evening News	Four Prestonians and one Victory machine
Sportsman	Two Victory and two Hoe machines
Sporting Life	Four Marinoni machines

Of the largely circulated weekly newspapers:

Lloyd's News	Six Hoe machines
Weekly Dispatch *Referee*	Six Marinoni machines
Weekly Times	Three Marinoni machines

90. An early folding machine by Harrild, *c.* 1879

91. The Cox-Duplex reel-fed flatbed machine as made by Dawson of Otley

14

REEL-FED FLATBEDS AND SHEET-FED ROTARIES

EARLY PRINTING PRESSES were not designed specially for newspaper printing which was simply part of the general trade. This situation changed when the circulations of major newspapers increased. This led to the development of multi-cylinder machines, then to type-revolving machines and finally to the web-fed rotary presses which utilized curved stereotyped plates.

But not all newspapers were of the size of London and New York daily papers and few in the British provinces and American small towns either needed or could afford stereotyping departments and large rotary presses. They could get by with the cylinder machine, or, in some cases, type-revolving machines, but tended to avoid the complicated business of setting type in specially curved chases. Type was printed from a flat bed for most kinds of printing and the small newspaper proprietor (who was often also a general printer) preferred to keep all his type composition that way. He was not, however, unaware of the benefits of printing several formes up together, or of feeding from the reel of paper; and if a machine could be marketed which was faster than the flatbed cylinder but cheaper than the rotary press it would be sure of a welcome.

The reel-fed flatbed perfecting press was the answer to the requirements of the smaller newspaper, and as far as can be judged the first of its kind was the *Duplex*, invented by Paul Cox in 1889 and built by the Duplex Printing Press Company, of Battle Creek, Michigan (Plate XLV). This web-fed machine had two flat beds, each of which could accommodate four-page formes, one above the other, and two travelling cylinders (reminiscent of the early cylinder machines), which printed an eight-page newspaper at each stroke. The web of paper stopped while being printed and then moved as the cylinders passed the ends of the type formes.

Competition with the *Duplex* in America was soon forthcoming when Walter Scott, of Plainfield, N.J., announced his version in 1891. This printed a four-, six-, or eight-page newspaper and delivered folded copies at a rate of 3,000 to 4,000 an hour. By 1898 the Campbell Company was advertising its *Multipress*, which it claimed could be operated by one pressman and a small boy at the rate of from 5,000 to 6,000 an hour. The machine could be had in two sizes—No. 14 (with a 22-inch page depth) and No. 15 (with a 23½-inch page depth). When the Campbell Company ceased trading the *Multipress* was taken over by the Machinery Trust, which eventually merged with the British Linotype Company to form Linotype and Machinery Ltd.

In 1903 the Duplex firm redesigned its press by adding a third bed of four pages, so that a twelve-page paper could be printed. To print an eight-page paper the top bed could be left idle. By 1906 the press could print four-, six-, eight-, ten- or twelve-page papers at 5,000 to 6,000 an hour, a rate not increased by 1920 when the *Duplex* 'Angle Bar' model was introduced.

Duplex's main competitor, however, was the Goss Printing Press Company of Chicago, by which it was eventually acquired in 1947. In 1908 Paul Cox transferred his affections to Goss, and they produced a reel-fed flatbed web perfecting machine from his patent of a somewhat different design to the earlier machine. In this new machine the two beds were positioned end to end and the cylinders reciprocated while revolving. It was built in two sizes—a two-page-wide version for an eight-page paper and one three pages wide for a twelve-page paper. The *Comet,* an eight-page flatbed perfecting press, similar to the first *Duplex,* was introduced in 1910 by Goss, which could produce papers at the rate of 3,500 an hour (Plate XLVI). The Goss *Cox-O-Type,* named after its designer, an improvement on the *Comet,* was brought out in 1928. It printed and delivered eight pages full-newspaper size or sixteen 'tabloid' size. An extra colour device could be attached.

It took only a few years for the *Duplex* to be introduced into Britain, and Dawson's, of Otley, began to manufacture it under the name of the *Cox-Duplex* in 1896. Their rivals, Payne & Sons, not to be outdone, announced their 'Flatbed Web Printing Machine' the same year ('four, six and eight page papers at 4,000/5,000 per hour'). There was also the *Lancashire* flatbed web perfecting and folding machine, made by T. Coulthard & Co., of Preston, in 1898. The first machine went to the *Carlisle Journal,* and others were installed at Grimsby, Oban, and Chatham. The *Northern Echo* (Darlington) and several South African and Australian papers also installed this machine. The manufacturers stated that it would print from 'Linotype, stereotype or ordinary type', but there was nothing special about this, as in all cases the formes would be flat. Speeds of from 3,500 to 6,000 an hour of four or eight pages were claimed.

Perhaps the best known reel-fed flatbed web perfecting machine in Britain was the *Cossar,* which was invented in about 1900 by Thomas Cossar, proprietor of the Govan Press, Glasgow, who succeeded in converting an old Dawson two-feeder machine for web feeding. The machine ran successfully for several years, but before long the firm of Payne & Sons decided to take up the manufacture, and Cossar was invited to go to Otley to supervise the building of the first machine. He did so, and stayed on to take charge of the design, construction and installation of all *Cossar* presses. The first complete machine was shipped to New Zealand in 1903, and consisted of a single-cylinder machine designed to print an eight-page newspaper in two operations. Two years later a two-cylinder *Cossar* two pages wide was installed in a Scottish newspaper office, where it remained in operation for many years.

About fifty units of this type of *Cossar* were sold before it was superseded by a new model in 1915. Soon afterwards, Payne & Sons became part of the newly formed firm of Dawson, Payne and Elliott. After that date the machine was continually improved, and the later versions were able to print up to thirty-two pages in one operation. There was also a limited second-colour attachment.

In the meantime, the *Duplex* was being manufactured in Switzerland by Buhler Brothers. It was basically the same as the American machine, although it printed on two strokes, and the folding mechanism was at the reel end. In January 1961 Gordon and Gotch of London bought the jigs, tools and manufacturing rights from Buhler's and made arrangements with George Mann & Co. to make the press in Leeds. By a series of amalgamations both Dawson, Payne and Elliott and George Mann became part of the R. W. Crabtree group of companies, owned by Vickers. In July 1968 an announcement was made that the group was to cease the manufacture of the *Cossar* press and concentrate on the *Duplex* machine which it built for Gordon and Gotch.

The group said it was obvious that they could not continue indefinitely to manufacture two machines which were basically similar in function and which competed for virtually the same market. An assessment of the two machines, based chiefly on comparative sales over a period of years, revealed that the *Duplex* had a better market potential.

The machine which started the trend towards reel-fed flatbed web perfecting machines in 1889 is thus, in the long run, by the accidents of history, the only survivor in its class.

There remains a final variation in the combinations of printing press feeding and printing—the sheet-fed rotary, the object of which is to produce high-quality work at rotary speeds. The process, if such it may be called, has had a chequered career, mainly due to the difficulties experienced in obtaining the requisite curved electroplates.

The sheet-fed rotary machine itself is hardly new, as, in effect, the first rotary type-revolving machines made by Applegath, Hoe and Marinoni, among others, were sheet fed. Others such as the *Whitefriars* could be either sheet or reel fed. The Hoe sheet-fed type-revolving book perfecting press of 1851 was an indication that such presses would be suitable for other than newspaper production, particularly if improved printing surfaces could be devised and automatic feeding apparatus attached. In 1882 Koenig and Bauer produced a sheet-fed rotary in Germany, and in America a number of manufacturers followed suit—Harris in 1897, the Scott *Flyer* of 1904, and particularly the inventions of Milton A. McKee, of the Cottrell Company, which made a speciality of this type of press.

For quality printing a better printing surface than the stereotype plate was required and the electrotyping process was considered. An electrotype is a duplicate plate made in a galvanic bath by precipitating copper on a matrix. Electrotyping as applied to type and illustration blocks came into use in the 1840s, although the deposit of one metal on another by galvanic means went back some forty years before then. The development of illustrated magazines in Britain and America encouraged electrotyping, but rotary printing required curved plates. McKee is credited with the perfection of the curved electrotype at the turn of the century and the *McKee* 33×46-inch multicolour sheet-fed rotary, with automatic delivery, was built in 1908 for use with his plates. Cottrell followed up with another in two sizes, 39×54 inches and 44×66 inches, in 1910.

Other American machines were the *Sterling*, produced in 1913 by the Printing Machinery Company, the Claybourn in 1930 and the *Kidder* in the same year. Such firms as Babcock and Miehle were also in the sheet-fed rotary field.

There was more incentive to build this type of press in the United States than in Britain, where curved electros were not very popular. From simple bending of finished electroplates, the Americans went on to produce them by a centrifugal casting machine, which backed up the pre-curved electro shell to the precise diameter of the press cylinder. This process enabled American printers to use rotary letterpress machines to print economically long-run quality magazines with colour illustrations. This was not the case in Britain where few machines were installed, and by 1958 Hazell, Watson and Viney were the only general printers producing

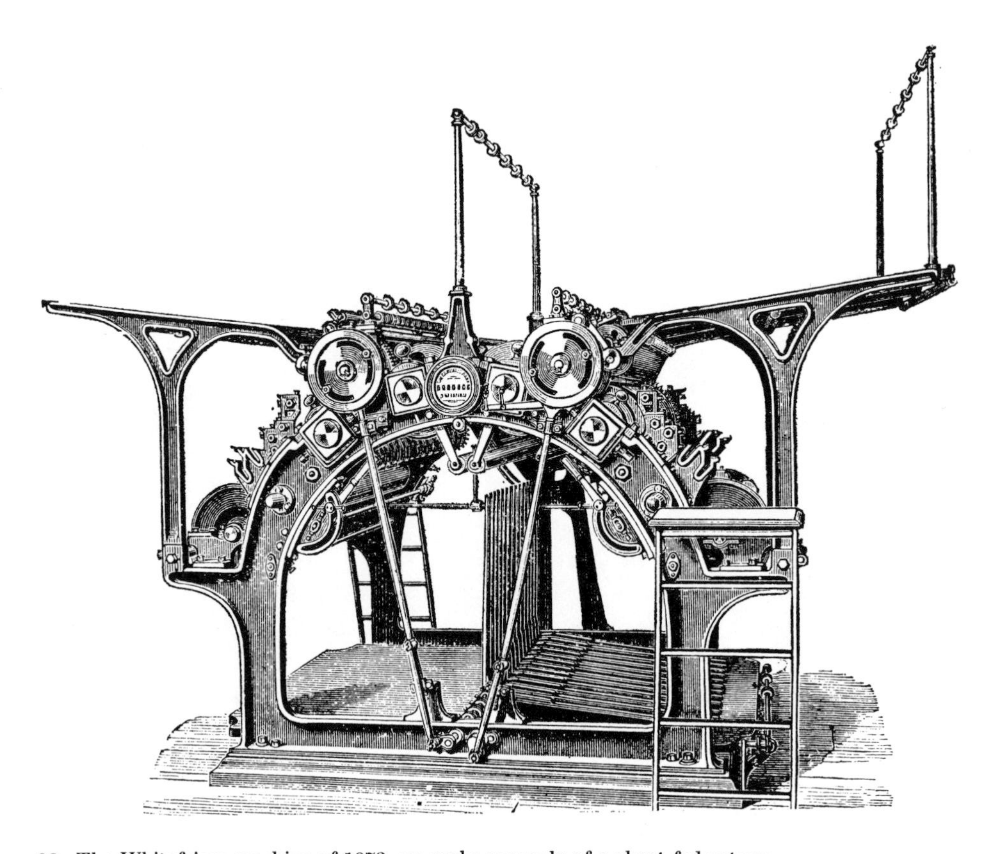

92. The Whitefriars machine of 1873, an early example of a sheet-fed rotary

high-quality printing from curved electros, but their battery of single-colour sheet-fed presses was installed to produce a specific magazine. Cold-curved plates presented them with difficulties, which were overcome when they installed the only centrifugal electrotype backing plant in Europe.

So that while a number of presses were manufactured for the production of bookwork from rubber and plastic plates, the development of modern sheet-fed rotary letterpress printing did not get under way in Europe until about 1957. The press manufacturers gradually became aware of the potential of the powderless etched original flexible plate. With the development of this process whereby thin, flexible 'wraparound' plates of either metal or of synthetic material became easily available the picture changed considerably.

The old-established firm of Dawson, Payne and Elliott designed a letterpress sheet-fed rotary to take advantage of this potential development in 1959. Also early in the field were Koenig and Bauer, with their *Rotafolio* machine. Heidelberg, and the American companies who had already been supplying the home market, also brought their machines to Europe.

This revival of interest in sheet-fed rotary printing led to a debate on the merits of rival printing processes, which still continues.

Research into the question of materials and methods of making flexible printing plates will also continue, but basically the principles by which the machine which, for the time being, carries out the printing are not new, and while, no doubt, refinements will be incorporated from time to time, the sheet-fed rotary like its companion the web-fed rotary has reached its zenith.

93. Hoe's double supplement machine (1882) was the first of its kind, and the first made with two printing sections at right angles

15

THE ROTARY PRESS FINAL STAGES

THE TENDENCY TO increase productivity on the printing press has always been twofold. Firstly, inventors and manufacturers worked to improve the efficiency of the working parts of the press, and secondly, to multiply the number of printing units until such time as a new and more advanced principle could be adopted. The development of the rotary press is no exception to this method of progression, and an endeavour will be made to trace the improvements which have taken place over half a century, and to indicate the ways in which the multiplication process has taken place.

Owing to the greater requirements of rotary-press building, there grew up a number of large manufacturing firms, with which the smaller engineers could not, and did not particularly want to, compete, and there was a decreasing need for publishers, such as those of *The Times*, to take a lead in actual manufacture. Before *The Times* in 1895 conceded that it was really no longer necessary for it to initiate new newspaper machinery, it had had its *Walter* presses modernized by Joseph Foster & Sons, of Preston. *The Times* management may have seen the light in 1894, for early in that year the proprietors of the *Daily News* had exchanged their *Walter* presses for those of a new design by Foster, which were among the earliest to foreshadow the way in which rotary-press installations would be made up in units. However, when *The Times* did change, it went over to Hoe presses.

A whole industry was thus at the disposal of the newspaper press, and much of the development resulted from an interplay of ideas between publishers and their pressmen and manufacturers and their technicians. For a time, and particularly with regard to ancillary processes such as stereotyping, individual newspaper printers continued to develop their own local techniques, but gradually it became commercially possible for outside supply houses to provide services.

In what was known as the 'wet flong' process the composition of the 'flong' was often a matter of individual choice. After a light rub over with an oily brush layers of paper (thin tissue, and sugar or blotting-paper) were placed on the face of the type, and pasted each to the other, with a mixture of flour, whiting and starch, or size, according to the particular views of the stereotyper. This, when thick enough, was beaten down into the spaces between words and lines with a bristle brush, so that the pasted paper (the 'flong') was indented by all the types. Sometimes a rolling press was used to press the flong into the type, instead of the brush. The forme, with the flong adhering, was put under the platen of a screw press, the underside of which was heated, and the forme was pressed well down into the face of the forme. The flong having dried into a matrix was lifted off and thus presented an accurate mould of the type. The matrix was brushed with French chalk and, being flexible, adapted itself for curvature in the semicircular casting box.

Various 'recipes' were available for making a flong. A. Eastwood, of Burnley, in his *A Short Treatise on Stereotyping* (1901) recommended for the paste a mixture of water, 'Spanish white', flour, farina and glue, and for the flong four sheets of 'best French tissue', two of mauve blotting-paper, one of ordinary red blotting-paper and one of brown paper for the backing. Various improvements in the wet-matrix process were patented, until in 1887 when George Eastwood in America and Hermann Schimansky in Germany began experiments which led to the dry matrix, which was finally launched in 1893. The dry matrix was a smooth, strong, pliable sheet capable of moulding, made to stand 700 degrees F. of heat.

Very quickly thereafter, in 1895, the Potter Company in the United States began to furnish dry matrices to their customers, but this system did not progress very rapidly. Not until 1908 did the London *Daily Mail* adopt the Padipp dry matrix, imported from Germany, although in the same year Joseph Dixon, of Liverpool, began manufacturing his 'Dixotype' matrices. Henry Wise Wood, of New York, saw the Padipp matrix being used at the *Daily Mail* and decided to encourage the dry-matrix process in the United States by securing the agency for the Padipp matrix. The dry flong, being much harder than the wet version, requires a powerful press to mould it into the forme of type. It is made from paper pulp—that is broken up fibrous material reconstituted with water—and made into sheets by the usual paper-making techniques but subjected to an intense drying process.

The wet-flong process continued in use on newspapers for a surprisingly long time, but eventually the dry matrix took over and the whole technique of making stereotype plates became more mechanized as more advanced equipment became available. When from 1860 onwards Dellagana in London, and Charles Craske in New York (whose experiments for the New York *Herald* went back to 1854) showed that stereotyping for rotary presses was possible, inventors began to think out the necessary equipment. James Wood, in London, invented an improved casting box in 1860, Dellagana a rolling press the next year and Alfred Vincent improvements in the flong in 1863; but the invention of the dry matrix and, in particular, the automatic casting mechanism—the *Autoplate*—invented by Henry Wise Wood in 1900, and installed initially at the New York *Herald*, led to the modern method of stereotyping on the major newspapers of the world.

The desideratum in rotary printing was an automatic press with as high an output as physically possible. This was achieved by degrees, leading to vast installations with the minimum of human control. In the process this led to changes in ink and paper manufacture and methods of supply.

The problem of set-off was tackled at first by special devices, but was solved eventually by inducing ink-makers to manufacture rapid-drying and non-set-off inks. Paper in rolls of uniform length and strength became possible when the paper-makers made a study of the problem, and the need for cheap newspapers stimulated the manufacture of 'newsprint', basically made from wood pulp.

The basic element in the rotary perfecting press is a unit composed of two pairs of cylinders, each with its own inking device. One of the pair is the plate or forme cylinder and the other is the impression cylinder. The expansion in output of the rotary press happened in a number of ways but really quite simply by increasing the number of these units and by adding to the number of plates attached to the plate cylinder. The manner in which the units were linked together made no difference to the principle of rotary printing.

E. L. Ford was the New York newspaper proprietor who tried to unite two or more printing mechanisms together, but, according to Hoe, achieved no lasting practical result, and it was not until 1882 that Hoe, after expensive experiments, constructed the *Double Supplement* machine, the first being installed at the New York *Herald.* The machine was constructed in two parts, the cylinders in one portion being twice the length of those in the other; the short cylinders being used for the supplements to the paper when it was desired to print more than eight pages.

The plates having been secured on the cylinders, the paper entered from the two rolls into the two parts of the machine, through each of which it was carried between the two pairs of type and impression cylinders, and printed on both sides. After this the two webs of paper passed over turning-bars by which they were manipulated one over the other and pasted together. The webs then passed down upon a triangular 'former' which folded them along the centre margin. They were then taken over a cylinder, for a final fold. At the same time a knife severed the sheet, and a rapidly revolving mechanism placed the newspapers upon travelling belts which conveyed them on for final distribution. The machine was claimed to turn out either four- to twelve-page papers at 24,000 an hour or sixteen-page papers at 12,000 an hour.

The Goss *Monitor,* first built in 1885, had a pair of three-plate-wide printing cylinders, enabling 18,000 four-page papers to be produced in an hour; and the same company's 'Straightline' of 1891 had two decks, two plates wide at the top and three plates wide on the lower deck, producing 25,000 four-page papers an hour.

In Germany Koenig and Bauer produced their twin rotary in 1890. This had two printing units arranged on both sides of the folder for the first time.

Hoe used the same ideas as in the *Double Supplement* for the *Quadruple* newspaper press in 1887. The supplement portion was increased in width and the press was claimed to be able to produce eight-page papers at a running speed of 48,000 per hour and ten- to sixteen-page papers at 24,000 an hour.

Another form of the Hoe *Double Supplement* and *Quadruple* machines, embodying substantially the same principles, was known as the *Straight-line* press. In this form of construction the cylinders were arranged in horizontal rows, or tiers, one above the other, there being two pairs of cylinders in each tier, with the folding and delivery apparatus at the end of the machine. For this press Hoe purchased patents taken out by Joseph L. Firm in 1889 which allowed the elimination of angle bars, thus reducing the strain on the web, and making web breaks less common.

This machine was thought to represent the ultimate in capacity for rotary presses, but in 1891 Hoe built the *Sextuple* machine for the New York *Herald,* in which the forme and impression cylinders were all placed parallel, instead of being at right angles as in the *Quadruple* press. It was fed from three reels and printed and folded 90,000 four-page papers in an hour. There were six plate cylinders, each carrying eight stereotype plates, and six impression cylinders. A double folder and two delivery outlets enabled the machine to cope with the output, and an automatic counter ensured that every fiftieth paper was pushed up an inch beyond the others.

On a more modest scale in Britain Foster's produced their 'patent triple web press' in 1894, which was installed at the *Daily News* and at the *Morning Post*. This was a twin machine—the 'offside' comprising two of the printing sections, with reels at each end, the paper passing between the impression cylinders to the centre of the framework, and there being turned over diagonal rollers to the centre of the 'near-side' portion, which comprised another printing section and the folding apparatus. By this arrangement each section was independent of the other, and allowed the printing of any size paper required. If a four-page paper was required the 'near-side' machine was sufficient; if an eight-page, the two sections were employed.

The development of the rotary press from the early part of the twentieth century was a matter of size and arrangement, of improvement to the ancillary service equipment, and of the addition of novel functions. As far as size was concerned, the Hoe *Sextuple* was succeeded in 1902 by the *Octuple*, which printed from four reels, each four pages wide, with a running speed of 96,000 for four-, six-, or eight-page papers. Presses grew in length and in the number of decks in which printing units were arranged horizontally. Plate cylinders were made to handle up to five pages across, and one or two plates around.

Manufacturers continued to supply smaller newspapers and printing houses with conventional rotary presses, but for the bigger houses installations were custom-built to meet the size and conditions of individual press rooms. Each manufacturer tended to use his own terminology and make his own claims; Goss, for example, stating in 1910 that its six-deck 'duodecuple' straightline press was the largest in the world. It was four pages wide, with twelve folders and a 42-reel paper magazine; maximum capacity was 48 pages at 25,000 an hour. Hoe's 24-cylinder multi-colour press was advertised in 1925 as the world's largest, but this has been passed many times since.

The *Philadelphia Bulletin*'s press room in 1956 was said to have the largest press line in the world. The equipment consisted of two lines of Hoe presses, each containing thirty-six units and seven folders with a total length of 319 feet. Presses were arranged in nine-unit groups, with four units ahead and five units behind each folder. At the same period it was also claimed that the Crabtree-built press room for Kemsley Newspapers, in Manchester, was the largest in the world.

The Goss Company is said to have introduced the first of the line type of press for the Chicago *Tribune* in 1920—a multiple low-construction press and folder unit in one line, all on one floor level. In effect, it was a collection of presses built on the unit principle in one row as against the deck principle of a number of presses one above the other. In 1927 the *Daily Mail*, in London, abandoned the deck type of press and introduced the unit type. This was produced in two forms, the *Unit* and the *Balcony*; the first with its bed plates on or about floor level and reels a floor

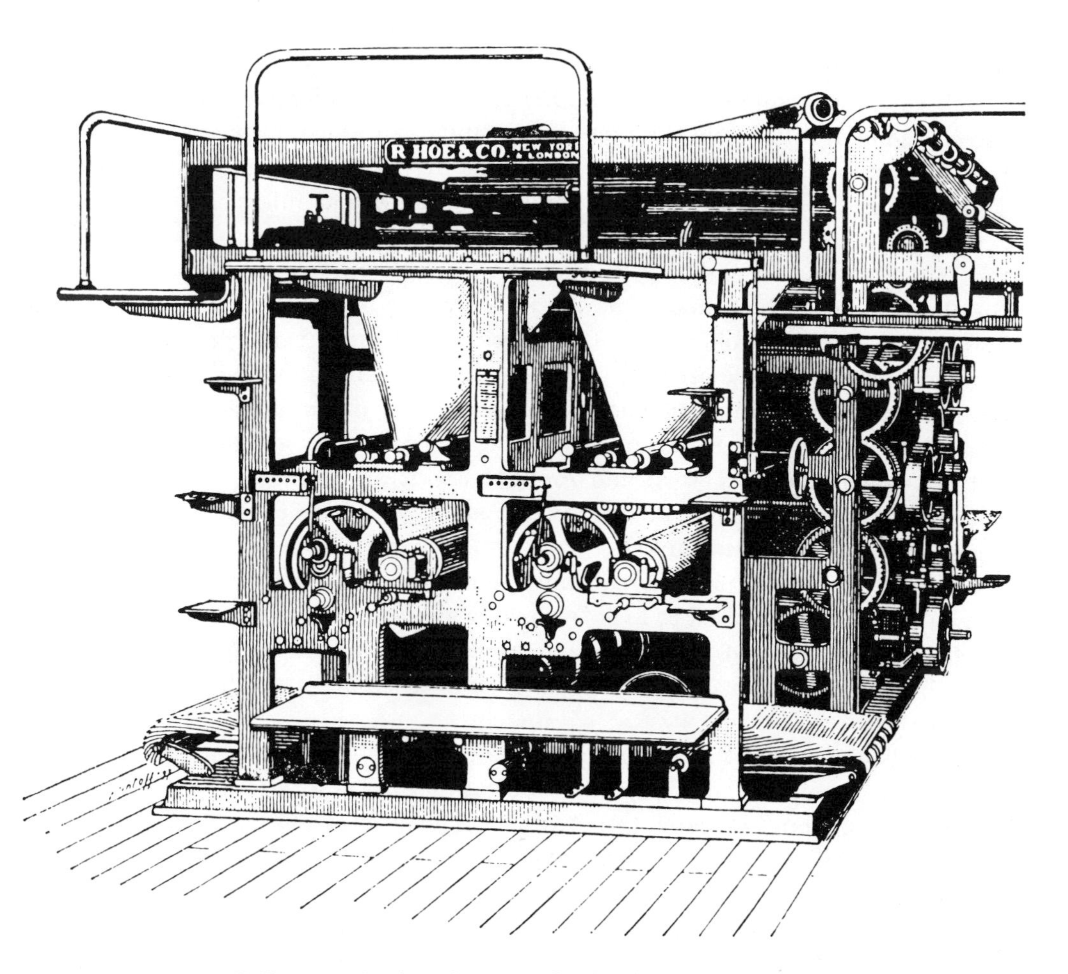

94. Delivery mechanism of the Hoe Quadruple rotary machine with two triangular formers. (Folded newspapers were delivered at the two stations near the floor)

below; the second with the reels at floor level with the bed plates some 6 to 8 feet above the floor on a balcony.

By 1933 both Hoe and Goss were enclosing the units and presses began to have a more streamlined look about them, and each part of the machine became increasingly more automatic in control. In the early days of the rotary the press had to be stopped if a reel of paper ran out. The core and shaft had to be removed, a new roll hung and the end of the expired roll spliced to it. By 1896 an American pressman, Irving Stone, invented a device which would bring a spare reel to the most convenient position.

Subsequent developments, both by British and American manufacturers, allowed the changing of rolls while the press was running, and for tension to be applied to the web automatically. During the 1920s the 'flying paster' came into operation. This allowed the splice to be made while the press was at full speed.

Originally, ink fountains were filled manually, and ink was conveyed from fountain roller to ductor and thence to distributor drums, roller and vibrators to the forme rollers which inked the plates. However, from 1915 Hoe replaced the fountain and ductor roller with a framework and pipes through which ink was pumped directly to the distributing system. The Hoe automatic ink pump by which the ink was pumped in a fine spray on to the inking drum, instead of being lifted from above the duct by means of a vibrating roller not only accelerated production but cut down laborious work and brought about an economy in ink and time. News ink is specially made, being thin in body and capable of flowing through the pipes, and is delivered in tankers direct to newspapers.

Curved stereotyped plates are mainly semi-cylindrical, and from 1908, when a special wrench was devised to lock two plates in one action to the cylinder, various mechanisms have been introduced to improve this aspect of rotary printing, including underside tension lockup, which pulls the plate more firmly against the cylinder.

As far back as 1866 the almost complete cylindrical stereotyped plate was suggested, but not until 1907 did this become a practical proposition with the *Duplex* tubular press. Plates were slid on to the cylinder from one end, leaving only a small non-printing gap, and permitting cylinders to be half the diameter of the semi-cylindrical plate type. Duplex made a four-plate-wide tubular press in 1914, and included this type of plate in its unit press in 1934. When Goss took over *Duplex* in 1947 it acquired the *Dek-A-Tube* single-width press, which it continued to build, but the tubular stereotype plate has never been very popular.

One meaning of the word 'fudge' is to make shift, to do work without proper appliances, or even to 'bodge up'. It became attached, in about 1889, to late-news printing devices and, hence, to the late-news item itself. Early attempts to introduce late news include Duncan and Wilson's patent of January 1879, which covered small cylinders to carry special movable type which could be slid into grooves. The type impressed against the impression cylinder and took its ink from the main supply. Mark Smith, of the *Manchester Guardian*, took out patents in 1886 for a small cylinder which was mounted on a shaft carried in bearings in the press frames with a type box and inking roller, which took its ink from one of the forme rollers.

95. Scott two-colour reel-fed rotary press, with folding mechanism, 1891

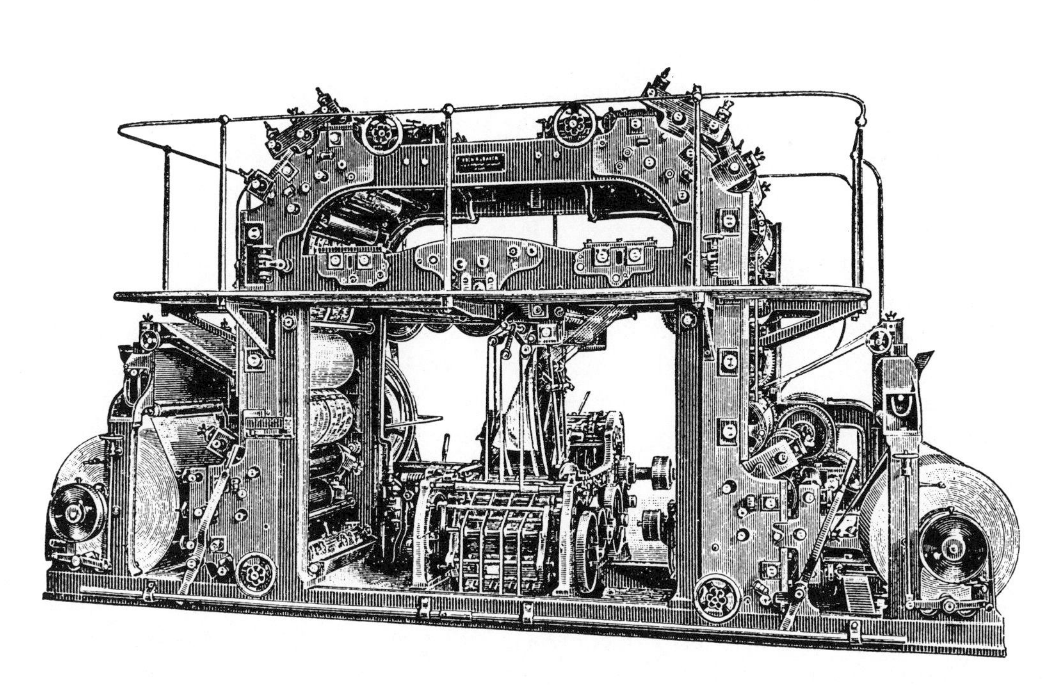

96. Koenig & Bauer twin rotary of 1890

The earliest 'fudge boxes' were simple and were arranged to take separate pieces of type in lines, but in 1888 John Petch and Thomas Mussel, of Middlesbrough-on-Tees, patented a means of rapidly inserting special matter in stereotype plates, on which Mark Smith improved by allowing for more boxes to be fitted to an auxiliary cylinder. The curved fudge boxes were arranged to take lines of type separated by brass rules. The type matter was inked by a separate roller, taking ink from the drum. Various improvements have taken place in the design of late-news devices from the end of the century, culminating in the self-contained apparatus, complete with its own ink fountain, inking rollers and separate impression cylinder.

This comparatively humble device was, however, among the reasons for the entry of a famous name into rotary-press manufacture. Charles Crabtree, of the firm of R. W. Crabtree, Leeds engineers, founded in 1895, was a friend of Mark Smith's son, Edgar Smith of the *Scotsman*, and in 1903 he improved on Smith senior's fudge box, taking out a patent in 1905. This work took Crabtree's into the field of printing and by 1914 they were about to produce their first rotary press. This had to be abandoned for war-work and it was not until 1919 that they were able to produce a press, the first of which (a three-deck type) was installed at the *Irish Independent*, Dublin, in 1921. Crabtree's purchased the London business of R. Hoe & Co., of New York, in 1938, and in 1965 became part of the Vickers group. But in October 1936 an unusual gesture had been made acknowledging the importance of newspaper-machine manufacture in London. Mansfield Street, Southwark, opposite the Hoe London works, was renamed 'Rotary Street'.

By 1911 fudge boxes were made to accept Linotype slugs instead of tapered type, and in 1927 Hoe's patented a duplex fudge and seal device which enabled late news to be printed in any column on the page including that directly under the coloured seal. The seal is a small block of colour used mostly as an edition indicator, and is cast in a special small box resembling that of the fudge. In 1928 Crabtree's installed at the London *Star* a non-stop late-news device which enabled fudge boxes to be changed while the press was running at speed.

The auxiliary cylinder for printing a second colour was common by the beginning of the twentieth century, but not until 1937, with the invention of the portable ink fountain by Auburn Taylor, an American pressman, did it become a practical proposition to print one or more colours on a rotary press at the same time as the black was printed. This process is known as R.O.P. (run of paper) colour.

The improvement of folding devices had been an almost continuous consideration since the rotary press got under way, but the folded papers still had to be carried away manually. From about 1912 when the American firm of Cutler-Hammer developed a newspaper conveyor, a number of 'creeping' devices by which newspapers are carried to the publishing room have increased in number.

With the coming of electronic controls, even fewer elements of the rotary press remained subject to manual intervention—ink control, register, colour register, and checking for web breaks, for example—and the most advanced rotary press is almost completely automatic in operation.

The rotary machine has thus revolutionized printing. Five hundred years after the invention of printing by Gutenberg, including four hundred years of printing at the rate of 250 single sheets an hour, rotary presses were turning out 70,000 newspapers an hour and handling up to 144 pages. By its development the rotary machine upset the careful labour structure of those four hundred years. Even the cylinder press did not disturb the skilled printer, who continued to perform his time-honoured mysteries of making ready. The rotary, which became a piece of precision engineering, was so different from the old machines that, at first, the man in charge was often a mechanic or engineer, who had perhaps helped to install it, and not a member of a printing craft union.

Indeed, the craft unions ignored the rotary press, and it was the displaced feeders and takers-away, none of them indentured printers, who began to man the new giants, and in time established a system of promotion by merit through various grades, and creating a new nomenclature, such as 'brake hand' (British) or 'tension man' (American) for the man who controlled the tension of the paper; oiler; fly-hand for the man who originally carried in stereo plates and cleared printed papers from the delivery end; 'reel hand' for the man who mounted the paper; and 'carrier-away'—one who washed rollers and helped the fly-hand. In the long run it was the carrier-away who lost his specialized job, as the conveyors increasingly did it for him, but the fly-hand still took the papers from the conveyors in the publishing room.

The craft unions then woke up to the significance of the rotary machines and began claiming that only apprenticed printers should have charge of them. This sowed the seeds of many later disputes.

It may well be that, like other printing devices, the rotary press has reached its technical and physical zenith, and that a complete change of principle in printing technique is due; but the signs are that for the large-edition newspaper, at least, the traditional letterpress rotary machine will continue in use for a long time.

Newspapers during the 1960s and early 1970s continued to install orthodox rotary presses. Typical examples on both sides of the Atlantic were the *News of the World*, London, which put in four Miehle–Goss–Dexter machines, capable of producing a 32-page broadsheet newspaper at a maximum speed of 70,000 copies an hour; and the *Toronto Star* which installed nine double width Viceroy Mk. II units, made by Hoe Crabtree Ltd., which can produce a 144-page paper, again at speeds of up to 70,000 copies an hour (Plate XLVIII).

L'IMPRIMERIE, descendant des Cieux, est accordée par Minerve *et* Mercure *à l'*Allemagne, *qui la présente à la* Hollande, *l'*Angleterre, *l'*Italie, *& la* France, *les quatre prémieres Nations chés les quelles ce bel Art fut adopté.*

16

APOTHEOSIS DEFERRED

AFTER FIVE HUNDRED years, the relief printing press has nearly had its day, but the signs are that it will be in use in various forms for another quarter of a century at least. During its five-hundred-year run it has developed from the very simplest device to a gigantic edifice, and has utilized virtually every known principle of motion.

Stamps with letters engraved in relief were used by the ancient Chinese and Romans, and it must have been a very simple intellectual step forward to think of stamping with individual relief letters, and Papillon and other writers have referred to the uniformity of some initials in European manuscripts as early as the ninth century, the conclusion being that this uniformity arose from the use of engraved stamps. The *Codex Argenteus*, possibly of sixth-century origin, in the university library, Uppsala, Sweden, seems to have been printed from such individual stamps—one letter after another.

So the first 'press' was the human arm, and devices for stamping individual letters and small messages continued to be made, as has been shown in this book, over the centuries and are used to this day in the ubiquitous 'rubber stamp'. When Gutenberg invented his sytem of casting types which could be assembled together for printing, a more advanced technique was needed, although the human arms were still paramount. Gutenberg rejected the lever and fulcrum in favour of the screw, which was to remain the basis of the printing press for four centuries. But later inventors went back to the lever and fulcrum and on this principle generations of amateur presses have depended.

The compound lever was used by nineteenth-century inventors before the adoption of that particular version called the knee lever, which became dominant before the cylinder press took over. Before that point, however, inventors had tried out inclined planes and wedges as the basis of the printing mechanism. The fly-wheel, windlass, the pulley, the spring and rack and pinion all played their part in the development of the press.

The coming of the cylinder machine, and then the rotary press, extended the ingenuity of engineers to the full, and many kinds of machinery were involved. Power of all kinds has been applied—man, animal (usually the horse), water, wind, steam, gas and electricity.

Men of many nations took part in the development of the press and it was a felicitous coincidence that two Germans, Koenig and Bauer, should have produced the first major advance since their fellow German, Gutenberg, invented the press. Because of Britain's more advanced economy at the time their work had to take place in London. Interest had been aroused on the Continent and at the beginning of 1816, for example, the Archdukes Johann and Ludwig came to England and made a special visit to see Koenig's machine, of which they had heard. It was demonstrated on 18 February 1816, the *Eleusinian Festival*, by Friedrich Schiller, being printed for them.

It is an odd thought that had it not been for the printer Bensley's limited outlook the first printing-machine factory in the world might have been in London instead of Oberzell, Bavaria. But, as it was, Koenig and Bauer had a hard time in Germany at first and Britain became the centre of printing-machine manufacture. After the war between the States, the United States took the lead, the expansionist conditions in that country providing much of the incentive to experiment and manufacture. Today, the major countries in printing-machine manufacture, besides Britain and the United States, are Germany, Italy, Sweden, France, Czechoslovakia, Sweden, Japan and Russia.

The printing press developed because education and literacy spread in the more advanced countries. Up to the beginning of the nineteenth century the wooden hand press was able to cope with the requirements of a limited reading public, although small improvements to the press were not precluded. The iron hand press was really an extension of these improvements, and it was not until the demands of the newly literate became insistent that a major change took place —the cylinder machine, leading in time, as the pressure grew, to the multi-unit rotary press.

While the smaller relief printing press will gradually be replaced commercially by mechanisms depending on photographic methods, the larger letterpress rotary will probably remain a feature of mass-production newspaper printing for some decades, because of its convenience and because nothing has yet appeared to challenge it for the combination of speed and output. Gradually, however, it will be replaced by presses based on some form of chemical process. Already, printing from metal type is on the wane. The next twenty years will see photo-mechanical methods of printing penetrating the last major stronghold of the relief printing press—the very large newspaper plants.

But for an unknown period production will be from some kind of press—that is to say a mechanism in which pressure is applied to varieties of printing plate. After that, the development will be away from pressure, as electrostatic and photocopying devices move from the small, office stage to larger-scale printing. When they take over Gutenberg's day will truly be done.

Until then it would be as well to defer the apotheosis of the printing press.

PLATES

I. Eighteenth-century English wooden press, now in the United States, and reputed to have been used by Benjamin Franklin in London, 1726

II. Close-up of box hose on an English wooden press

III. Improved press of Anisson-Duperron, Paris, 1783

IV. Ramage press No. 913

V. Ramage press built about 1820 with iron platen

VI. Representation of the improved press of Wilhelm Haas, Basle, 1772

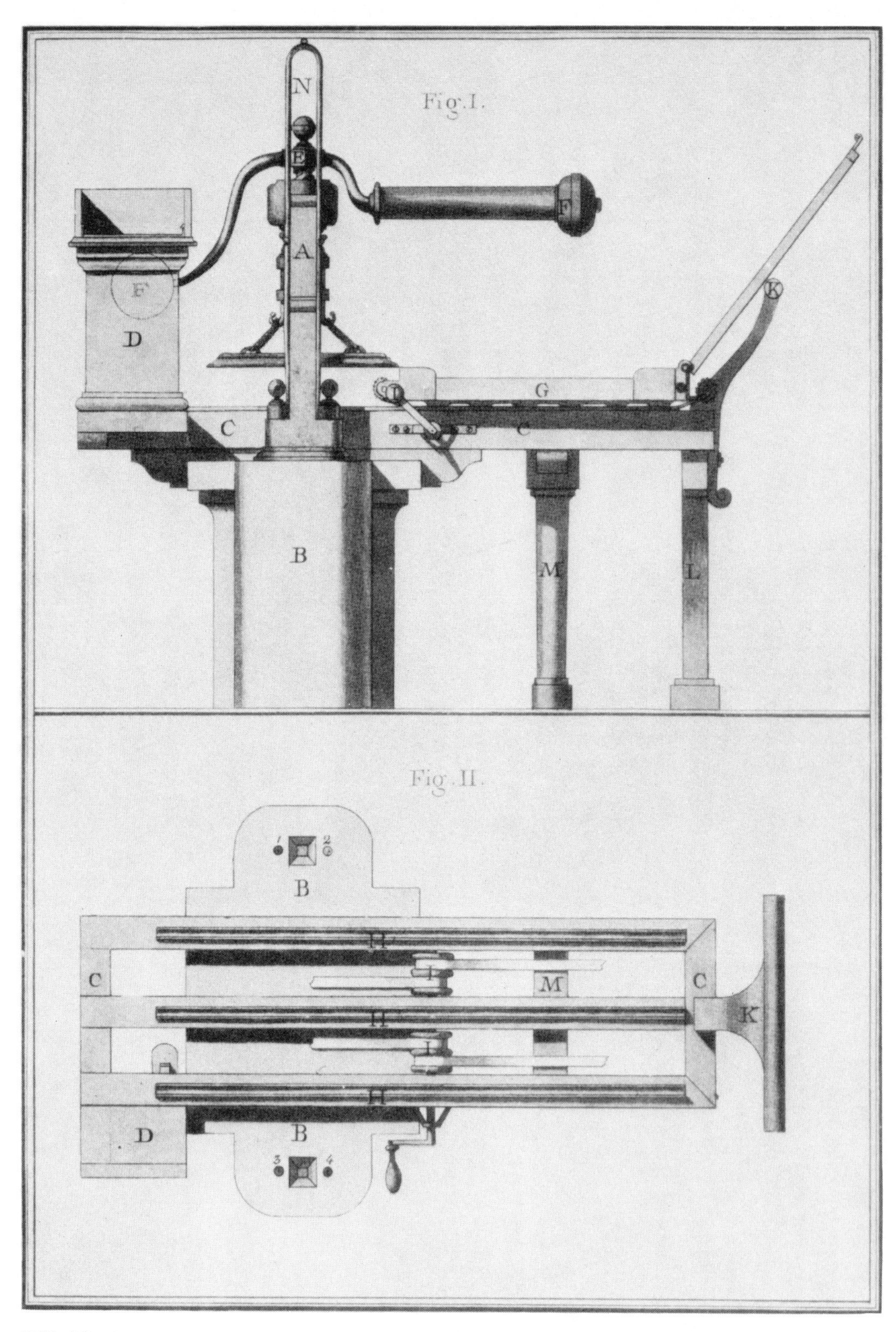

VII. The Haas press. Improvements relate to iron frame (A), stone base (B), combined fly and bar, pivoted at (E) with weighted ends (F, F) and iron ball rack (N). Fig. II shows no change, in principle, from the wooden press

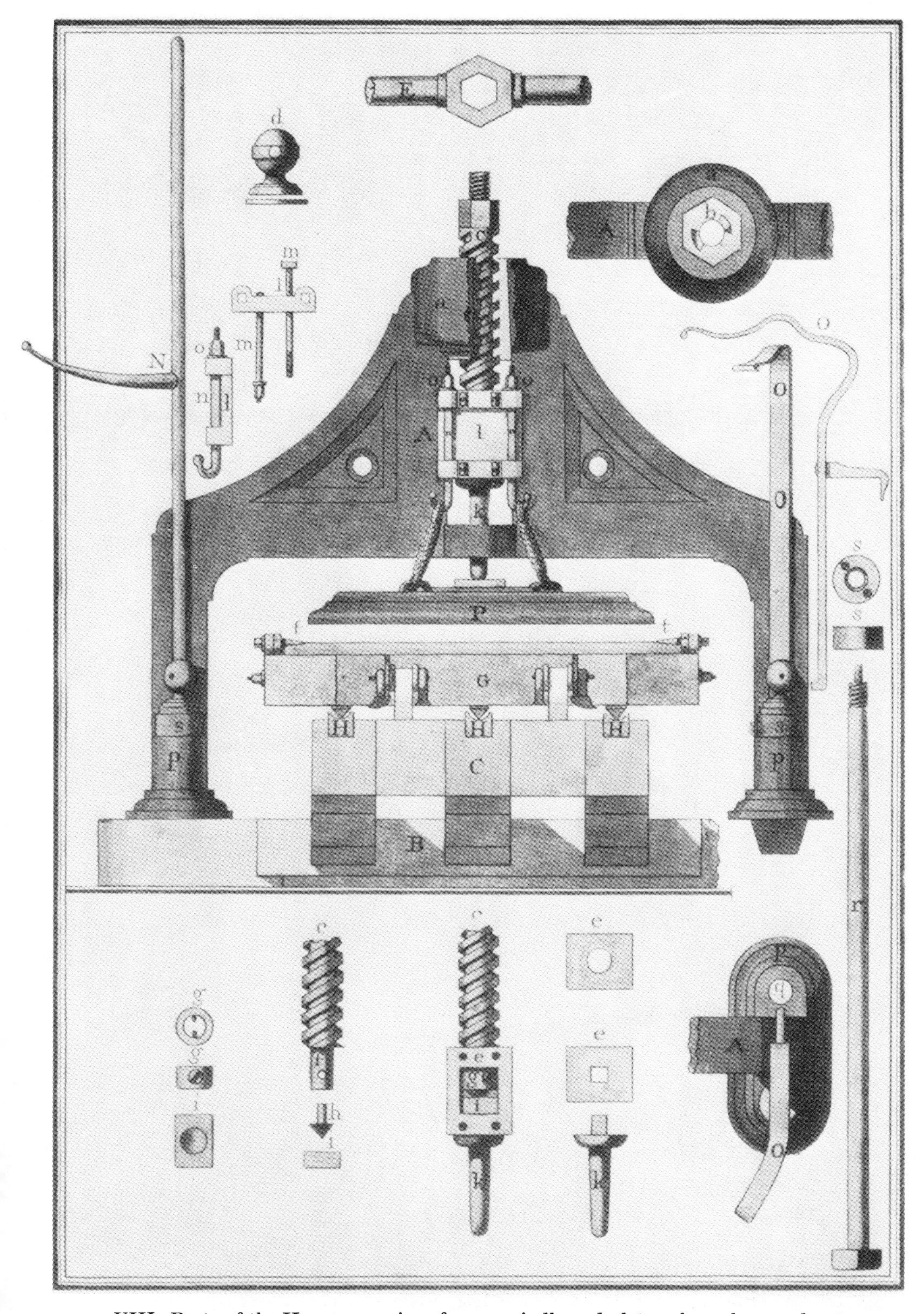

VIII. Parts of the Haas press—iron frame, spindle and platen; brass hose and nut

IX. Stanhope press No. 9

X. Stanhope press No. 67

XI. Tissier Stanhope press, 1847

XII. Munktell Stanhope press

XIII. Gaveaux Stanhope press (Bishop Pompallier's press, N.Z.)

XIV. Clymer's Columbian press No. 13, 1818

XV. Columbian press by Thomas Long, Edinburgh

XVI. Columbian press by Fr. Vieweg, Brunswick

XVII. Columbian press by H. P. Hotz, The Hague

XVIII. Britannia press by R. Porter, Leeds

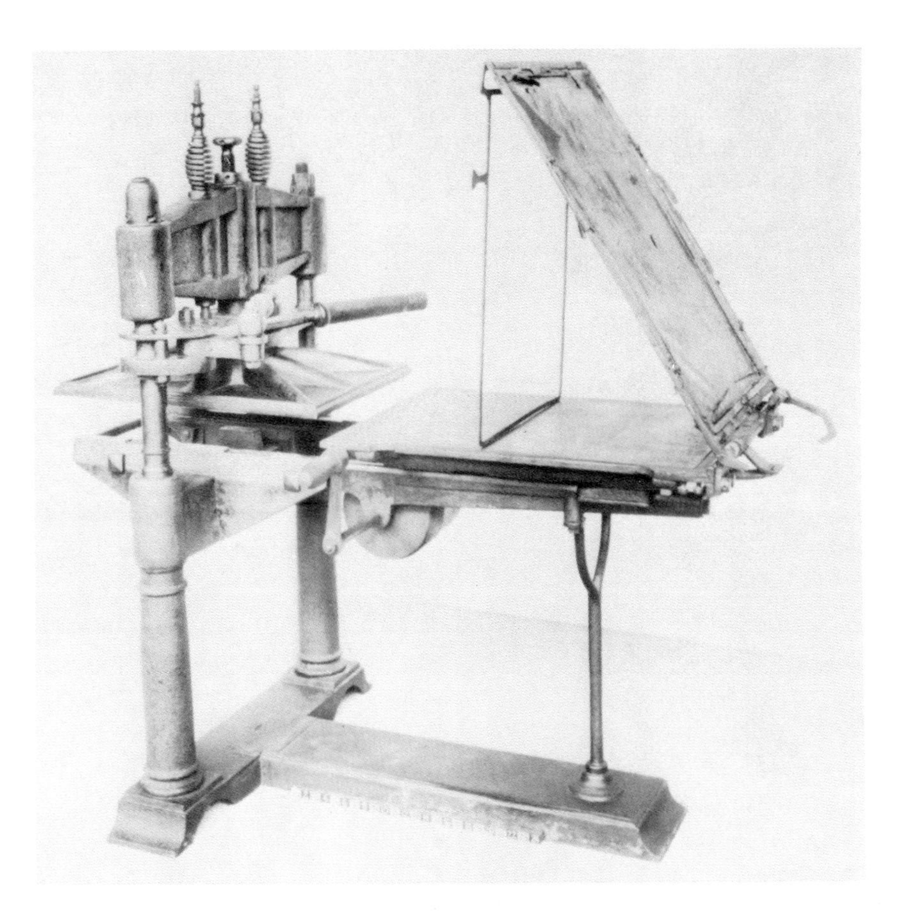

XIX. Cogger press

XX. John Lyons press, Mullingar

XXI. Wells press (above left)

XXII. Hoe-built Stansbury press (above)

XXIII*a*. Moffat press. Moffat's portrait (in right-hand corner of illustration)

XXIII*b*. Smith press

XXIV. Stansbury press

XXV. Albion press (Cope's executors)

XXVI. Atlas press

XXVII. Albion press in Singapore

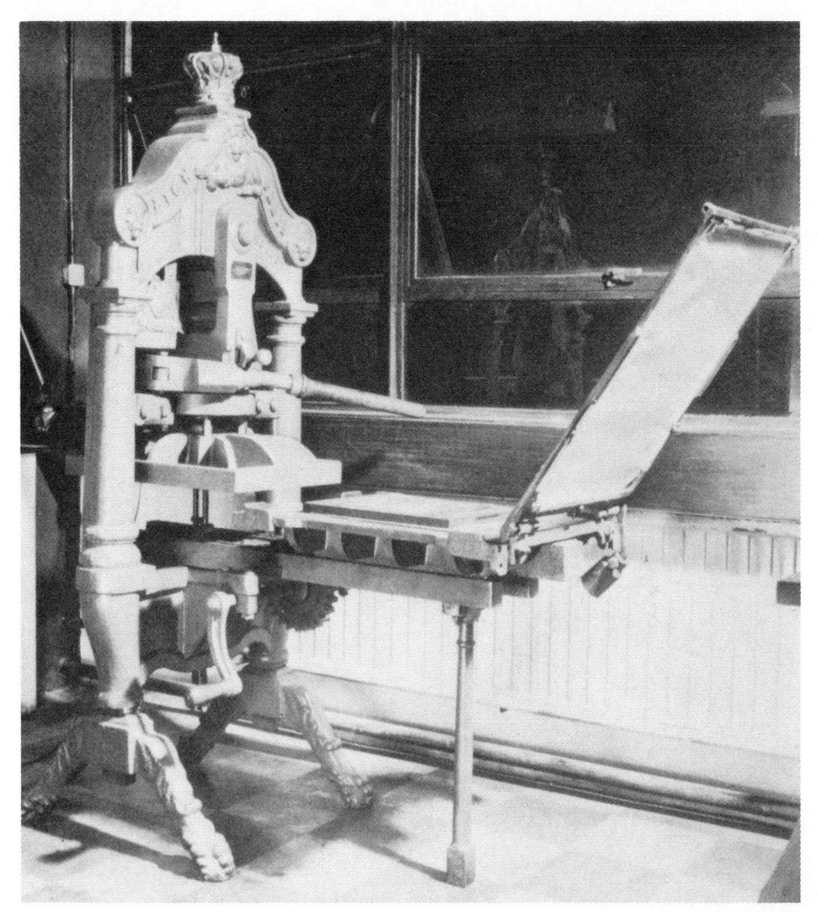

XXVIII. Lion press

XXIX. Harrild Herculean Albion press

XXX. Albion press by Frederick Ullmer

XXXI. Washington press, with medallions, by Palmer & Rey, San Francisco

When I had drawne these Engines before passed knowing it to bee a necessary work, to bee my
[illegible] of such worthy marciall men, as haue occasion sundry times, to put such deuices in
practize, for the benefitt of England, and their own renowne, I thought good by your
Honours fauour to present sundry of these vnto such, yt by their profound Judgements
and long experience, might giue their censure, whether I be worthy for my guiftes from
God to bee maintyned in my princes seruice and natiue country, or for want thereof, to
seeke maintenaunce abrode in forrein nations, as heetherto I haue bin forced to do. but
for that to draw and wright any number by hand, it wolde bee a tedious work
Conferring with sundry printers to the end I might print a certen number of them
but none woulde compounde, vnlesse they might reape what I had sowen, and for
my corne to giue mee chaffe. Wherefore I deuised this kinde of presse, wch is portable hauing the force of 24 men, to the end I might print them with myne owne
handes, for such as are worthy to possesse them, and will not imploy any Inuention
contayned herein otherwise then lawfully. The mouements of this presse are soe
plaine to bee perceiued, that it needeth no declaration.

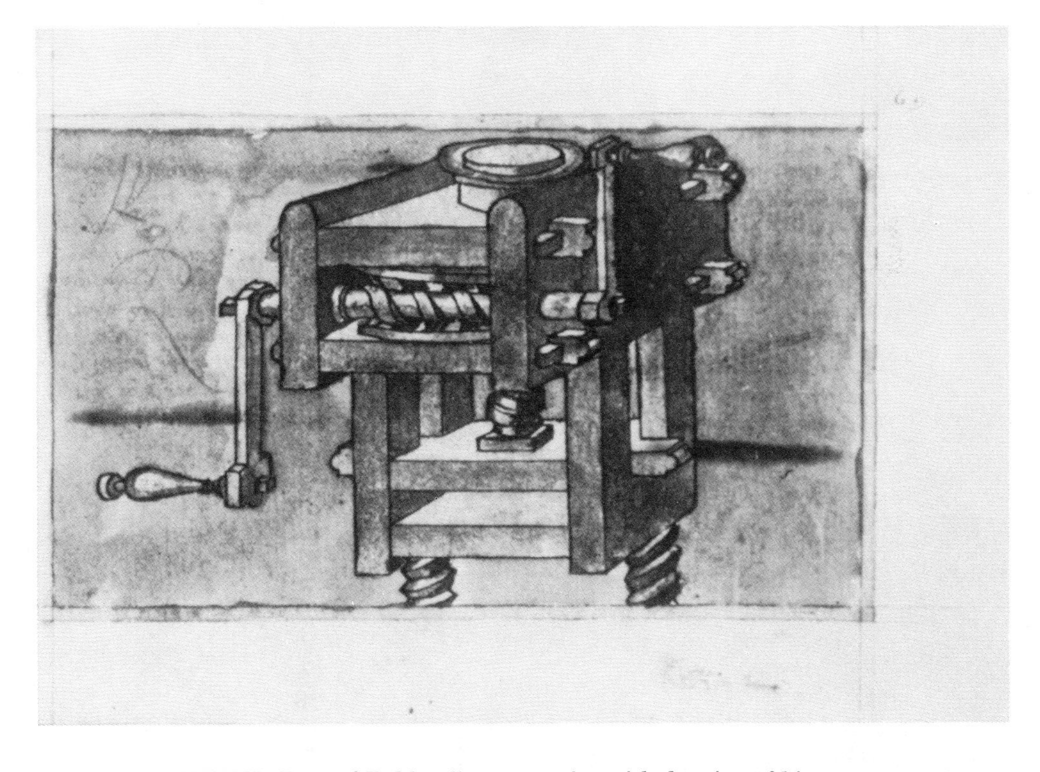

XXXII. Page of Rabbard's manuscript with drawing of his press

XXXIII. Model of Koenig's 'Suhl' platen machine

XXXIV. Model of Koenig's first cylinder machine, 1812

XXXV. Model of Koenig's double machine for *The Times*, 1814

XXXVI. Model of Koenig's perfecting machine, 1814

XXXVII. Contemporary photograph of 'Our Own Kind' machine of William Dawson, *c.* 1850

XXXVIII. Prouty country newspaper press, *c.* 1887, nicknamed 'Grasshopper'. Simple travelling cylinder type

XXXIX. The Alligator press

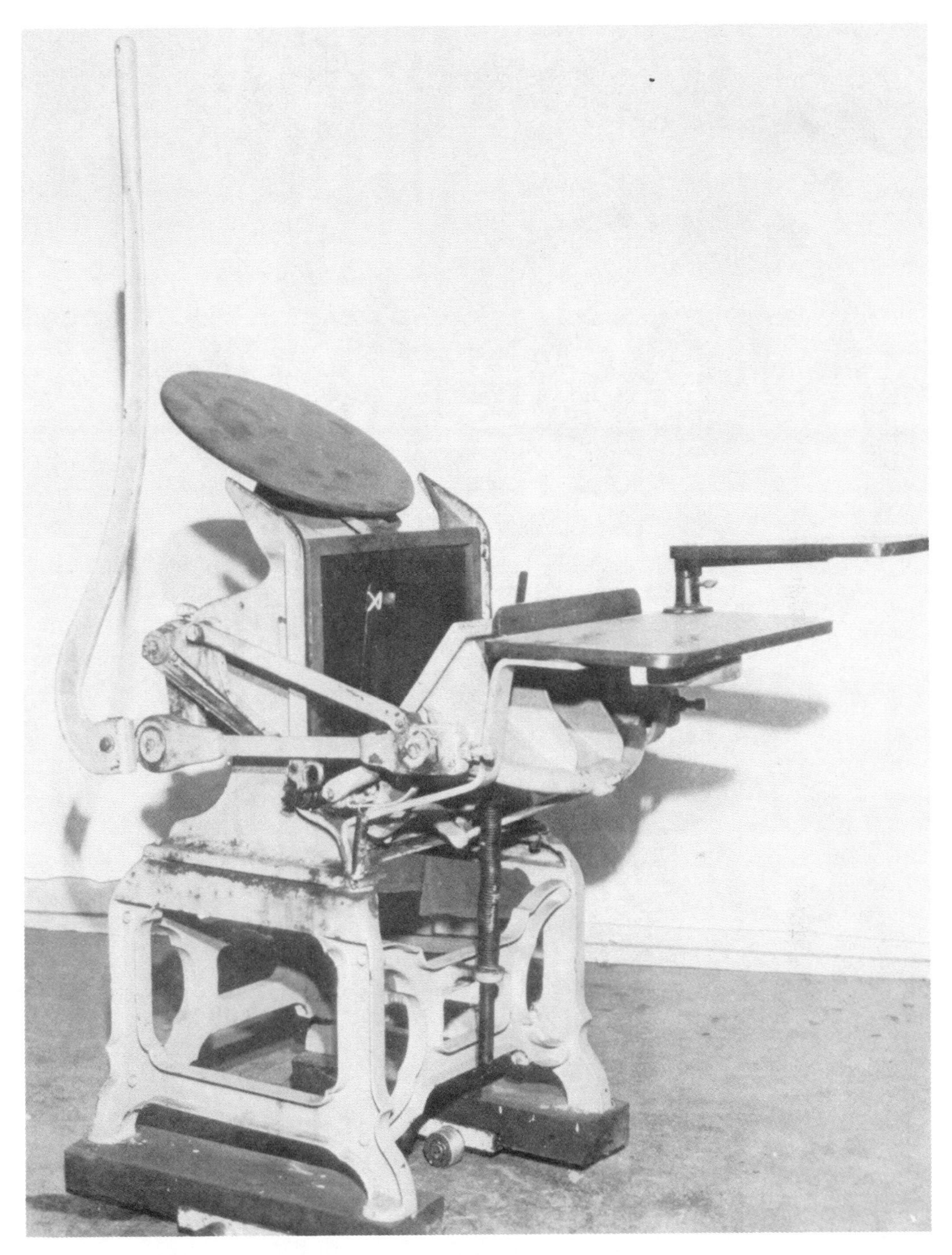

XL. Unidentified jobbing platen, 66 inches high, last used by Oklahoma Indians to print newspaper, *The Falling Leaf*

XLI. Columbian job platen, *c.* 1880, made by Curtis & Mitchell, Boston. Misleadingly called a 'rotary press', it had no connection with Clymer's Columbian hand press

XLII. Thomas Nelson's rotary machine

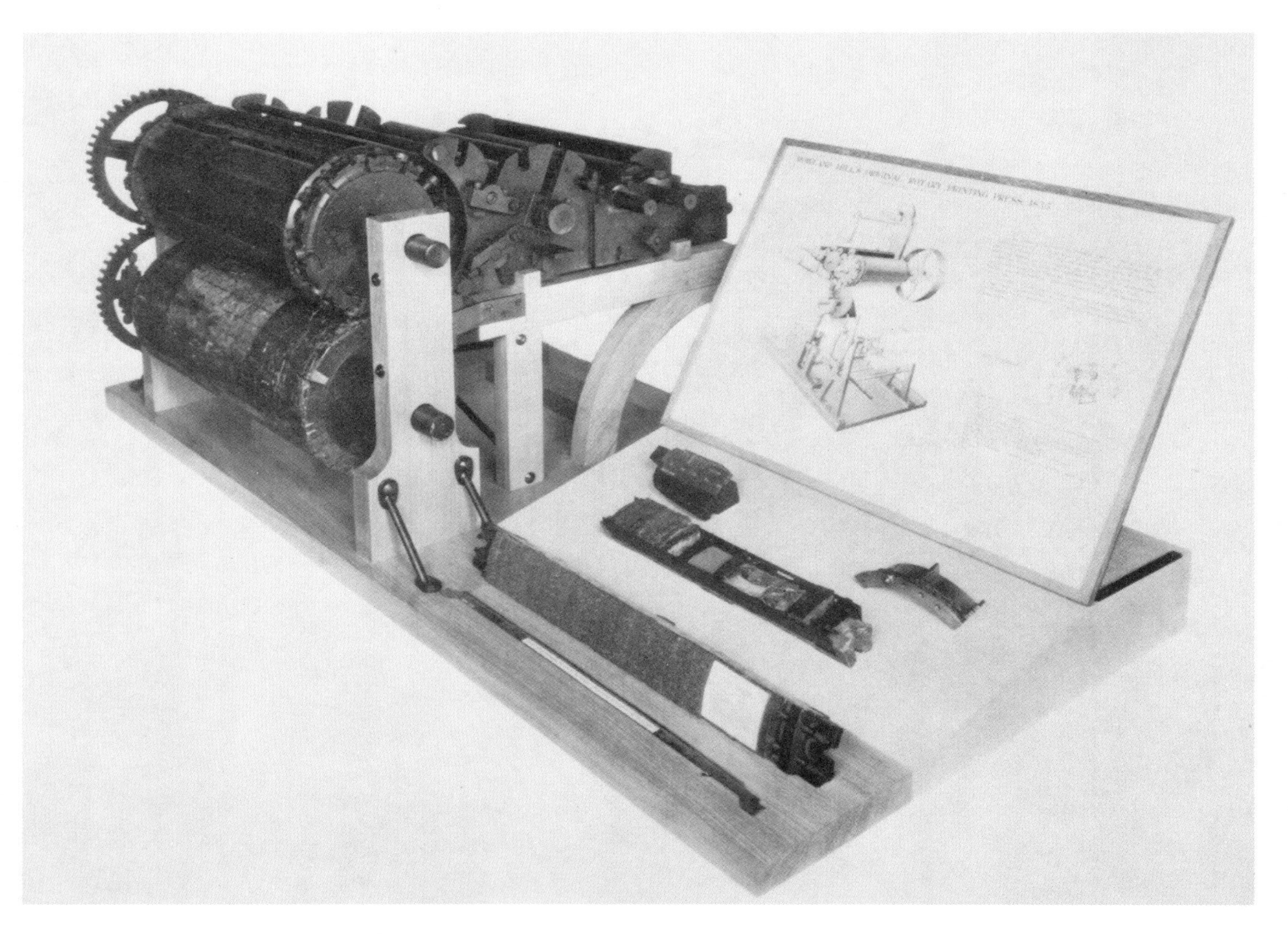

XLIII. Elements of Rowland Hill's rotary machine, 1835

XLIV. Contemporary photograph of the Walter rotary machine, *c.* 1870

XLV. The Duplex reel-fed flatbed machine, *c.* 1889

XLVI. The Goss Comet reel-fed flatbed machine, *c.* 1910

XLVII. Hoe rotary press used by Alfred Harmsworth (Lord Northcliffe) to print *Answers* (founded 1888)

XLVIII. One of five nine-unit Viceroy Mark II rotary presses made for the *Toronto Star* by Hoe-Crabtree Ltd.

XLIX. Scaled-down wooden press reputed to have been used by François de Fénelon

L. Table model Stanhope press 'La Typote'

LI. Side view of 'La Typote'

LII. Golding Official press, 3×5-inch chase, *c.* 1885

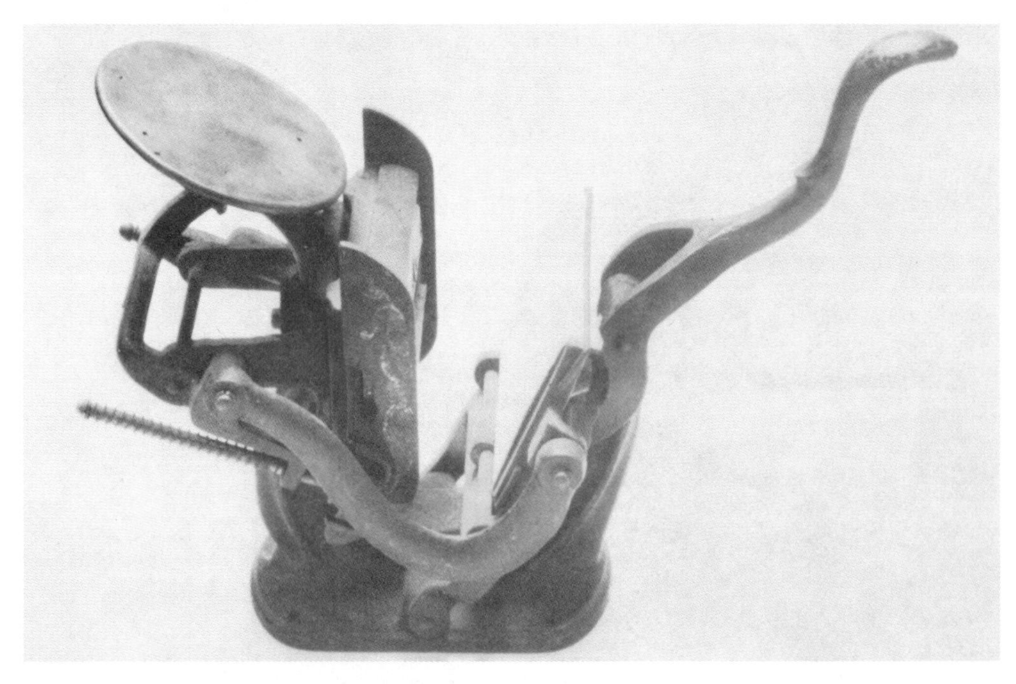

LIII. Golding Official press, Junior model, $2\frac{1}{2} \times 3\frac{1}{2}$-inch chase, *c.* 1880

LIV. No. 1 Giant, $2 \times 3\frac{3}{4}$-inch chase, patent 1876

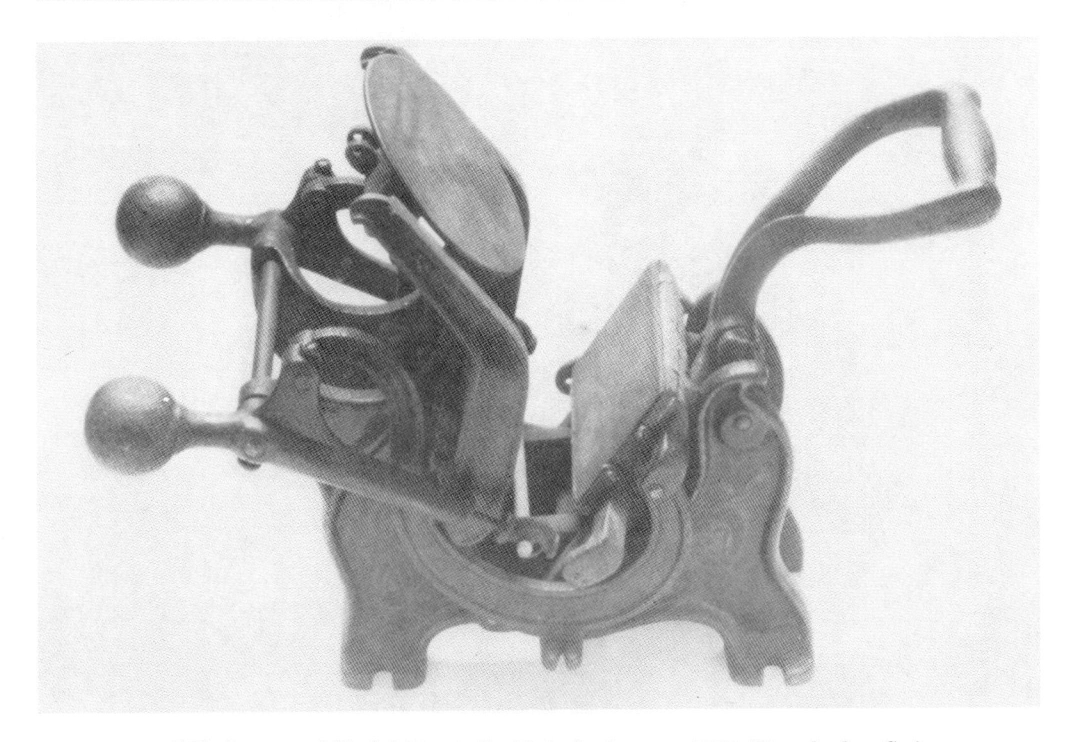

LV. Improved Model No. 1, $5 \times 7\frac{1}{2}$-inch chase, *c.* 1870 (Doughaday Co.)

LVI. The original flatbed Adana press, 1923

LVII. The Baby Adana flatbed press, 1928

LVIII. No. 1 High speed Adana platen press in sheet metal (wooden handle), 1938

LIX. No. 1. High speed Adana platen press in cast iron and Mazak, 1933

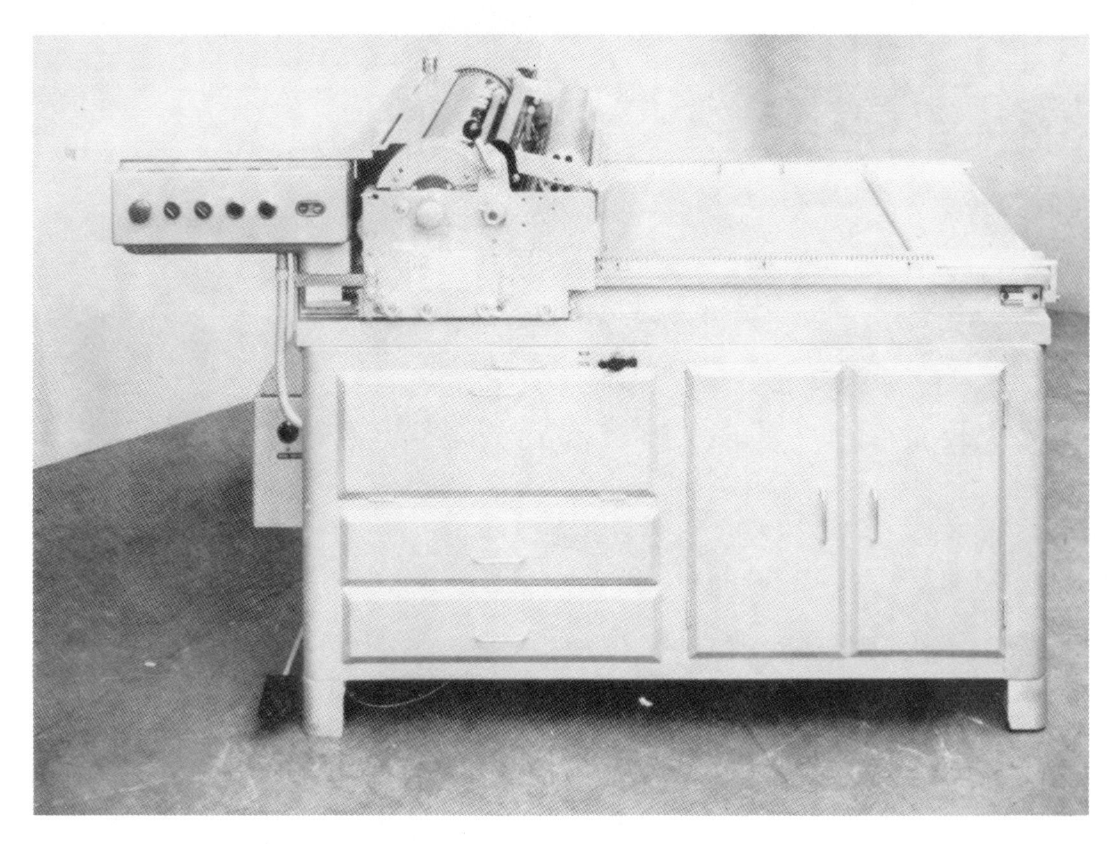

LX. Western proofing press

LXI. First Vandercook proof press—rocker type

LXII. Vandercook Universal four-colour wet-on-wet proof press

LXIII. Hoe galley proof press, *c.* 1850

LXIV. Hoe improved galley proof press, *c.* 1870, made for 'Doctor' Miles

APPENDIX

98. Miniature Albion press exhibited by Mrs. Daniel Jones at the 1862 Exhibition

APPENDIX I

MINIATURE, 'TOY', AMATEUR AND CARD PRESSES

SINCE SMALL-SCALE AND what are generally thought of as amateur printing presses are derived from different classes of press it is considered more appropriate to deal with them in an Appendix, rather than in a chapter placed in chronological order. These presses range from what would be considered today as a mere stamp for marking to precision-built jobbing platens, and a certain confusion in nomenclature has to be faced, since it is difficult to draw a distinction at times between, say, a 'toy' and a 'model' press and between an amateur and a professional apparatus. A straightforward model—that is to say a representation—is not a press if the intention is that it should not actually print.

It is unlikely that the miniature common press (scale one-sixth) made in 1812 by Henry Thomas Ellacombe, on show in the Science Museum, London, was made for practical use; but there are other such small presses on which it is perfectly possible to print. Whether some were intended for amateurs or were salesmen's models it has so far been impossible to determine.

Even intention is not necessarily a guide. The word 'toy', in the sense of a plaything, has been used since the sixteenth century, and from the early part of the nineteenth century the word has been applied to models of engines of industry and war, no matter how large originally—everything from railway trains to cannon. Such toys can be made on any scale, and can incorporate various degrees of realism. Toy presses have been made, and although not necessarily very sturdy, have managed to perform the function of printing. While intended primarily for children's play they can have a more serious function.

A shire-reeve in Iceland, one Sigurdur Petursson, used a 'toy' press in the late eighteenth century for his own purposes, but in another Danish dependency, Greenland, the German missionary Jesper Brodersen produced a little hymn-book in 1793 on a small 'handbuchdruckerey', acquired on a visit to his homeland, thus becoming Greenland's first printer. The first printing in Mississippi was carried out in 1798 by Andrew Marschalk on a small portable mahogany press he had acquired in England.

The line between a professional and an amateur press is hard to draw when it comes to small items of printing. The same kind of small hand press can be used by a tradesman to print visiting cards for profit and by a private individual to print his own poems for pleasure. A child's toy press can become a weapon of war when used to print propaganda behind enemy lines.

The earliest printers printed either for a limited scholarly audience or for the rudimentary reading public, but whatever their motive, their equipment was much the same, including full-scale wooden presses. No particular changes were made, as far as we know, when royalty and the nobility felt the urge to print. Some amateur printers, such as the young Louis XIV or the Duke of Cumberland and his sisters at St. James's Palace in 1713, may have actually worked a press, but others, Horace Walpole at the Strawberry Hill Press or Sir Egerton Brydges at the Lee Priory Press, left the printing to professionals, but their presses were the same as those used in commercial printing houses. The tradition continued with the great private presses of the nineteenth and twentieth centuries—the Kelmscott, the Doves and the Ashendene, which used full-scale iron hand presses.

Nevertheless, from the beginning of the sixteenth century, joiners and smiths exhibited their skills by constructing scaled-down wooden presses, which were used by members of the French aristocracy, including Madame de Pompadour. A number of beautifully made examples survive in France—in Paris at the Musée du Conservatoire National des Arts et Métiers, in the Collection Dangon and in the possession of Madame Darangués; in Lille at the Musée de Lille; and that in the Musée de l'Imprimerie et de la Banque at Lyons traditionally is said to have been used by François de Fénelon (1651–1715), French prelate and author, but no authenticating documents exist (Plate XLIX). M. André Jammes, of Paris, has a model of Genard's improved press of 1787 (see Chapter 2). In Brussels there survives a small travelling press of the orthodox type but with shortened legs for use on a table, but a portable press in the museum at Hamm, Westphalia, can be closed up into a box and rests on a chest of drawers.

Pierre-Simon Fournier le jeune (1712–68), the French typefounder, was, by special decree, allowed to set up a little printing office where he had a small press, so it can be presumed that scaled-down versions of the wooden press were made to order for working purposes. In the Caxton Celebration Exhibition Catalogue of 1877 it is noted that Mr. John Coe lent a 'toy press used by Charles I'. If this were genuine it would probably have been a model of the usual wooden press of the day.

99. Berri's People's Press. (See pages 238 and 240)

The sixteenth-century military adventurer Ralph Rabbards, whose manuscript is in the Osborn collection, Yale, did, however, suggest a change, although whether it represented a technical step forward is doubtful. He had found the cost of having his words printed too high, and so, he wrote: 'Wherefore I devised this kinde of presse, which is portable having the force of 24 men, to the ende that I might print them with myne owne handes, for such as are worthy to possesse them, and will not employ any Invention contayned herein, otherwise than lawfully. The movements of this presse are soe plaine to be perceived, that it needeth no declaration.' It is a pity that Rabbards did not describe the workings as, despite his words, his drawing is not entirely clear. There is a conventional screw terminating in a fixed platen, beneath which is a stationary bed. Where the till would normally be there is a slow-acting worm, with a handle protruding from the side of the press. Presumably if this were wound the worm would drive the screw and hence the platen downward with, as Rabbards suggested, considerable force. If this press had been built it would have suffered from the drawbacks of other screw presses—it would have been slow, and might have been so powerful as to crush the type.

Whether Rabbards had his press built or not, it indicates that even for portable presses the average inventor did not think of abandoning the screw, and when Adam Ramage, in the United States, towards the end of the eighteenth century began to build lightweight portable presses they followed the same principles as those of the larger size.

The most primitive way of obtaining an impression on paper from a piece of inked type is to press it on with the fingers, which according to Edward Rowe Mores, in his *A Dissertation on English Typographical Founders and Founderies* (1778), was the habit of Mr. Passmore Stevens, who printed imaginary title-pages by using the ball of his thumb as an ink pad for small work, and a rolling-pin for larger. Dr. C. H. O. Daniel, whose private press is among the best known of the nineteenth century, began as a schoolboy by using his thumb as an ink-ball and press, before graduating to a toy press and then to an *Albion* hand press.

Mores, himself, owned a small press, known to the trade as a bellows press, which reverted to the pre-Gutenberg lever and fulcrum principle, and which, as has been noted, Isaac More and William Pine used for their projected 'Leaver and printing press' in 1770. The bellows press derived its name from the familiar household instrument, and consisted of two thick boards, one of which had a recess for inserting type, hinged together, and terminating in two handles. The pressure applied by grasping the handles is sufficient to print from a few lines of type to produce small cards and labels. There are two surviving presses of this kind in the St. Bride Institute collection.

James Boswell records in his *Life of Dr Samuel Johnson* that on Saturday, 23 March 1776, on a visit to Lichfield 'We went and viewed the museum of Mr. Richard Green, apothecary here, who told me he was proud of being a relation of Dr Johnson's. It was, truely, a wonderful collection, both of antiquities and natural curiosities, and ingenious works of art. He had all the articles accurately arranged, with their names upon labels, printed at his own little press; and on the staircase leading to it was a board with the names of the contributors marked in gold letters. A printed catalogue of the collection was to be had at a bookseller's.' Green's press was probably of the bellows kind and used for the printing of the labels, while the catalogue would have been produced on a full-scale press at a local printing office.

That portable presses on the bellows principle were capable of printing more than a couple of lines is indicated by a letter of the Italian poet and dramatist Vittorio Alfieri of 1786, when he referred to a little portable press on which he could print a sonnet of fourteen lines. An even more ambitious press of this kind must have been that mentioned in the Börsenverein *Katalog der Bibliothek* (Leipzig, 1885–1902). The reference was to a pamphlet, produced by a Hamburg printer in 1772, which described 'pocket printing offices', and was itself 'printed with the pocket printing office'. What precisely was meant by 'pocket' is left to the imagination, but a very small printing press in a box which could go in a fairly capacious pocket was sold in the 1920s by the British firm, Adana, for 5*s*. 6*d*.

By the middle of the eighteenth century printing was being taken up as a hobby by fashionable people, and a scaled-down screw and platen press perhaps being too expensive to produce, the older and simpler idea of lever and fulcrum was resorted to. A man who seems to have thought of this even earlier than Moore and Pine was a London engraver, J. Sutter, whose presses must have been moderately well known as they are referred to by Rowe Mores, and the inference is that they were of the bellows type. A Sutter advertisement appeared in the *Reading Mercury and Oxford Gazette* of 30 October 1769, as follows:

TO NOBLEMEN, GENTLEMEN and LADIES curious in PRINTING

J. SUTTER, Engraver and Printer to their Royal Highnesses George Prince of Wales and Prince Frederick,

St. Martin's Church-Yard, St. Martin's Lane, London,

HAS invented Portable PRINTING PRESSES on a new construction; by which two folio pages may be printed, more or less as required, with the greatest expedition and exactness.—The young Nobility and Gentry of both sexes, by means of this excellent invention, may be easily brought acquainted with the Poets and other Authors, by reprinting the beauties of each and thereby strengthen the memory and early improve the understanding.—Types, and the rest of the Printing Material with proper instructions and examples, how to compose the Types, and cast off the Copies, may be had at a small expence, on the shortest notice.—Small Printing Presses, the size of a pocket volume, are just now invented, by which Cards, &c. may be printed.

Sutter, 'the Original Inventor', also advertised 'Stamps and Liquid' for marking linen. The instrument for holding the stamps was so constructed that any person might mark linen or books 'with all expedition'. Complete alphabets with sets of figures were available as were 'Cyphers, Crests, Coats of Arms, &c' executed in wood or metal.

The fame of this invention must have spread abroad, for an 'English' pocket printing office was advertised for sale by the Breitkopf typefoundry of Leipzig in the 1780s for marking washing; and the Marquis de Bercy in 1791 wrote of two friends receiving little English printing sets as New Year gifts.

It was a way of printing slightly in advance of Mr. Passmore Stevens's thumb. A block or relief letters, held in a holder, inked, usually on a pad, is stamped downwards by hand. This technique goes back to antiquity when seals were thus imprinted. English monarchs had their signatures printed in this way. Edward I is known to have had a stamp of this nature, and the signature of Henry VI (1436) cut on a wooden stamp survives in the Public Record Office, London. The *Sutter*, *Breitkopf*, Holtzappfel *Monotype* and *Baum* 'printing press' (mentioned later) were instruments for holding type, and in the United States the *Gorham* 'press', advertised in 1875 as a 'card press and linen marker', utilized special short type for this purpose.

Once the industrial uses of rubber were discovered in the middle of the nineteenth century the way was open for the production of rubber stamps and rubber types, but metal type continued to be used for stamping small pieces of information. From 1852 a number of inventors patented a 'self-inking stamping apparatus'. One of these devices, patented by E. E. Bartlett, of Salem, Massachusetts, on 29 December 1857, survives in the Martin Speckter collection in New York. Type locks upside down in a drum, which is pushed down by a blow of the hand and, pressure being removed, it is forced back to its original position by springs, but in so doing permits an inking roller to traverse the face of the type to allow the next stamping to make an imprint.

Such were devices for printing a very small amount of type, but if more than Alfieri's fourteen lines of type were to be printed, greater power than that produced by squeezing by hand was required, and Sutter's press for two folio pages must have incorporated some device to increase the power, and can hardly pass muster as a 'bellows' press, which term should be restricted to those where pressure is applied by two handles (hence a 'pair of bellows' is used to describe the domestic implement). It should not be used, it is suggested, of those stronger, hinged presses which utilize leverage to procure pressure.

Philippe-Denis Pierres, whose attempt at an improved common press has been noted, produced two versions of a hinged press from 1786 onwards, which, although effective, did not as yet use a lever and fulcrum on top of the platen. Plans are available of Pierres's presses, and two models exist in the Musée du Conservatoire National des Arts et Métiers in Paris. Plans, models and references have been described in some detail by David Chambers in the *Journal of the Printing Historical Society*, No. 3 (1967). Pierres also had a rival, M. Beaucher (or Baucher), who produced presses of this type. In Pierres's first model the platen is held with a frame, hinged to one end of the main construction, and which is able to move up and down within the frame, extending from which are iron rods carrying a cylindrical counterweight. Bolted across the middle of the platen is an iron bar, from each end of which hangs a hook. From a shaft running under the press two hook-shaped catches rise at each side. Connected with the shaft is the operating lever, extending beyond the shaft and ending in a spherical weight.

When the platen is folded on to the forme the catches at the sides of the press automatically engage themselves in the platen hooks and the lever is then depressed. After impression, the lever is raised, assisted by the weight at its end, and the pressure taken off the catches and hooks. On the catches is a forked handle, and on this being pulled the platen opens, again assisted by a counterweight.

One of the disadvantages of the hinged press is that pressure is greatest at the end where bed and platen join, but Pierres overcame this drawback by making the platen free to move up and down so that the paper did not come into contact with the forme until pressure could be applied uniformly over the sheet. As far as can be gathered, Beaucher's press differed from that of Pierres in that its spherical counterweights were violent in action (a failing of the later *Stafford* press), and its platen was not free and met the forme at an angle. Beaucher's beam, however, in the shape of an inverted arc, pressed on the platen in the middle, which may have assisted the distribution of pressure, an ever-existing problem for inventors.

The beam on Pierres's press was across the width of the platen and was pulled down by the hooks and catches. That on the *Ruthven*, described earlier, was similarly pulled down by clamps. These were not quite the same as the beams of antiquity.

100. The Stafford press

Somewhat nearer was that on the press devised by J. Stafford, of Bingham, Nottinghamshire, early in the nineteenth century. His press had a horizontally hinged platen, over which was a forged beam, which was pressed down with a screw at one end. Very heavy counterweights were needed to raise the combined tympan and platen, and Hansard warned: 'If the tympans and platten chance to be thrown up quickly a concussion will be produced such as would endanger the floors of many places used as printing offices.' Hansard provides an illustration of the *Stafford* press, an example of which survives in the St. Bride Institute collection. It was not a commercial success, for apart from the dangerous counterweights, it needed a separate movement to screw down the beam. What was needed was a lever which would apply pressure with a single movement.

In the middle of the nineteenth century others lower down the social scale than Sutter's 'Noblemen and Gentlemen' took up printing as a hobby, and engineering firms began to realize there was a market for small presses. These at first took the form of scaled-down versions of existing presses. Sir Alexander Boswell, in 1815, had purchased a portable *Ruthven* press in Edinburgh, exchanging it later for a full-sized version, so that manufacturers were already aware of a demand. Holtzappfel & Co., a well-known engineering firm, in 1834 included small presses 'on the

Stanhope and other principles' in their stock and T. Cobb made a small *Columbian* for printing visiting cards. No small surviving *Columbian* has yet been traced but two table-model presses of the *Stanhope* type exist—one in Paris and one owned by Muir Dawson in Los Angeles, which has a nameplate 'La Typote', but no other identifying marks. It was bought in France and it is fairly certainly of French manufacture. Height is $26\frac{3}{4}$ inches, length (full out) 52 inches and width 20 inches. Platen size is $11\frac{3}{4} \times 17\frac{3}{4}$ inches.

It is strange to think that one of Britain's greatest engineers, Sir Joseph Whitworth, may have been connected with the manufacture of miniature presses while working at Holtzappfel's and before setting up on his own in Manchester, where, amidst the manufacture of cannon, he managed to build the giant Hoe ten-feeder printing machines for *The Times* in 1858. Ten years after his death in 1887 his firm became Armstrong Whitworth & Co., now part of the Vickers group, which has strong printing-press manufacturing interests.

Examples do survive of one scaled-down iron hand press—the *Albion*. 'Card' sizes of this press became popular not only among amateurs but also among those trade printers who produced small items. At the Exhibition of 1862 Mrs. Daniel Jones showed a small *Albion*, which, from the illustration in *Cassell's Illustrated Exhibitor* looks as if it was a royal octavo or 'card' size. George Bullen, of the West Central Letter Foundry, Judd Street, Brunswick Square, marketed Bullen's Improved *Albion* in this size for five guineas, and one similar, bearing the plate of T. Matthews, of Smithfield, survives. Charles Morton, of City Road, made small *Albions* on orthodox lines, but his *Cosmopolitan* press, which deviated somewhat from the *Albion* principle, did not meet with the approval of *The Bazaar* magazine, the findings of which will be mentioned later.

A surviving Lockett & Sons list (no date) advertises a size even smaller than royal octavo ($10\frac{1}{2} \times 6\frac{1}{2}$ inches) and that was the *Amateur* (7×5 inches). Hopkinson and Cope also supplied this size *Albion*, and another between royal octavo and foolscap octavo (the smallest used by the average printer), namely one with a platen of $14\frac{7}{8} \times 9\frac{3}{8}$ inches, made to operate on a stand 26 inches high.

Two further miniature *Albion* presses are known to have been manufactured, but by whom is a mystery. One unique survivor in the St. Bride Institute collection, which came from East Anglia, has a platen of $6 \times 4\frac{1}{8}$ inches, and is only $18\frac{1}{2}$ inches high. While it is possible to print with this press, it is not certain whether it was made for amateur use or whether it was a salesman's model. The other kind, of which two specimens are known, consists of a table *Albion* without a rounce mechanism. The forme has to be pushed under the platen, as presumably was the case with the very first printing presses. There is no spring in the cap and the platen is raised by a flat spring in a box at the base of the piston, working on lugs inside the two cheeks. No clue has been found to indicate where these presses were manufactured. The first (in a private collection) has no lettering on it; the second (in the Hertford Museum) has 'Albion press' cast in the head. The platen size is $6 \times 4\frac{1}{2}$ inches. Paul Shniedewend, of Chicago, also abandoned the rounce mechanism for his rather larger *Baby* (10×8 inches) (see Appendix II—The Proof Press).

Nineteenth-century manufacturers must have found that scaled-down iron screw and bar presses were too expensive for the lower-grade market, and, like their predecessors of the previous century, they resorted to the simpler lever and fulcrum press.

In 1839 Holtzappfel, which had made small-scale *Stanhope* presses, began to manufacture Cowper's *Parlour* press, and although no copy of the first set of instructions seems to have survived, the third edition of the manual, *Printing Apparatus for the Use of Amateurs* (1846), contains descriptions of three sizes of Cowper's press and of Holtzappfel's *Monotype* printing press, a device for lettering plans or producing small labels, using one type at a time, as the name suggests.

101. The Parlour press

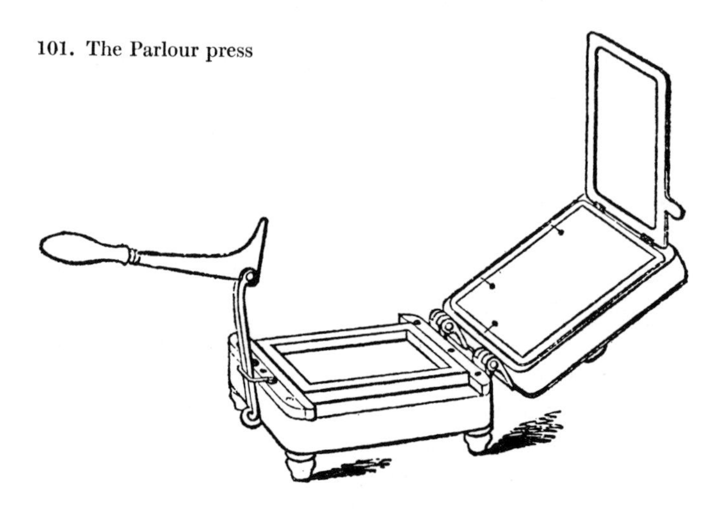

The *Parlour* press was the invention of Edward Cowper, brother-in-law of Augustus Applegath, who built the biggest printing presses of the day. For his *Parlour* press Cowper used a genuine captive lever working on a fulcrum, and from the Introduction to the Holtzappfel Manual it can be ascertained that Cowper's invention owed its origin to 'a small and old instrument, known amongst printers as the *Bellows Press*. . . .'

The Introduction continues: 'A very superior modification of this press was constructed in metal, by the celebrated Timothy Bramah, for printing the dates and consecutive numbers on the Bank of England Notes, after the general impressions had been printed from a copper plate.' It was, in fact, Joseph Bramah (1748–1814), of Pimlico, inventor of a hydraulic paper press in 1790 and a papermaking machine in 1805, who the next year patented a simple press for those not acquainted with printing. A variant of this apparatus, with a series of wheels with raised numbers, was adopted by the Bank of England for numbering bank-notes. The platen was screwed to a lever, with handle attached, and the already engraved bank-note was brought down into contact with the raised numbers. The lever, however, moved on an axis, on which was fixed a catch or click. This acted in the teeth of a wheel when the lever was thrown back as far as it would go, moving the wheel round one tooth and consequently another wheel to which the numbering types were fixed, thus bringing up another type to be printed.

Whether Bramah actually produced a version for amateurs is not known, but Cowper could have seen the numbering machine when visiting the Bank to discuss the production of counterfeit-proof notes. The Holtzappfel booklet points out the disadvantage of the simple hinged press, in which, while pressure is sufficient near the hinge, is weak at the other end as the amount of pressure continually decreases from hinge to handle. Cowper, it is claimed, overcame this problem—'Mr Cowper conceived the idea of the removal of the former handles of the bellows press altogether, and producing the pressure at the open end, by a modification of the well-known mechanical combination, the *toggle* or *knee joint*.' The pressure in the *Parlour* press at one end was due to the hinges and at the other to the toggle-joint.

Cowper might not have been completely original in his approach. His lever looks much like that of Bramah, and there is also a surviving wooden *Parlour*-type press in the Salford Art Gallery and Museum, with a lever pivoted at its open end, which increases in power as it descends in the same way as Cowper's. While it is not known exactly when this press was made it is conjectured that it dates from the eighteenth century.

The smaller models of the Cowper *Parlour* press printed a page 6×7 inches, but a foolscap folio size, with a bed 15×10 inches, a folio demy size, with a bed of 20×13 inches, and a broadside foolscap size, with a bed 20×13 inches, were also available. Only one model of the small size is known to have survived.

During the 1870s, Jabez Francis, of Rochford, Essex, made what appeared to be a version of the *Parlour* press in cast iron, but the principle of working was not the same. The lever action raised the bed from below by acting upon two interlocking U-shaped bars, and the press was therefore not strictly on the bellows principle. In a way, it was a miniature 'bed and platen' press. The production cost 25*s.* for the 6×4 inches size; 3 guineas for the $9\frac{1}{2}$×8 inches size and 5 guineas for the 5×9 inches size. One of these decorative presses survives, but it is more than possible that Cowper and Francis presses exist in various parts of the country as

102. Bramah's press for numbering bank-notes

they must have been made in considerable numbers. Holtzappfel were still advertising 'printing presses (for regimental & parlour use)' as late as 1927.

The Holtzappfel booklet also referred to a 'hand chase'. This held a few lines of type, not exceeding three inches in length and one in breadth, and was for the printing of small labels, in the manner of a hand stamp. After the type was set, it was locked in securely with a wedge. The chase was supplied in a box with roller, ink and inking tray, and a cushion on which to lay the paper to be printed.

More ingenious was the *Monotype* press, contrived by Charles Holtzappfel, for printing labels, only one copy of which was required. This avoided the setting up of type, as the device permitted the printing of one type sort at a time through a kind of carbon paper. The press consisted of a sliding table on which was laid paper and carbon paper on top. A piece of type was held in a notch at the end of an arm and was pressed down by means of a lever. When each letter was printed it was necessary to move table and paper to the left to make room for the next letter.

A press which excites interest, but about which there seems to be no further information, is described by the anonymous author of *The History of Printing*, published by the Society for Promoting Christian Knowledge, in about 1862.

Amplifying a footnote about the bellows press, the author wrote as follows: 'There is for instance the "bellows" press, a primitive invention quite different in principle from the presses in general use. In this simple machine the bed of the press is stationary, neither rolling in nor out—the platen is fixed at the back of the tympan, and the impression is produced by forcing the wet sheet, tympan, platen and all, in a rather sudden and violent manner upon the inked face of the type. Of course it is a very inferior printing that is produced by such a machine. There is, however, a modification of the above simple instrument in the shape of a small iron press, which will print anything not larger than a foolscap folio in a creditable manner, with a very fair impression. In this the iron platen at the back of the tympan somewhat resembles the platen of the *Albion* press: it is not thrown violently down upon the type which rests on the stationary bed, but is laid down gently, and at the moment of contact is caught at its left-hand extremity by a projecting tongue or iron communicating with a metal plate on the floor; on this plate the workman places his left foot, and treading with his whole weight, thus produces the impression which is generally sharp and clear, though it may be rendered very bad by careless handling. The advantage of using this kind of press lies in the rapidity with which it may be worked—the writer has often worked five hundred copies an hour with it single-handed—on the other hand its disadvantages are many: it will not impart pressure enough for anything but small jobs, and unless humoured in the handling the impressions will be slurred.'

Unfortunately, the author does not give the name of the press, if it had one, or that of the manufacturer, and no press fitting the description exactly has been traced. One which has some resemblance was made by Christian Dingler, of Zweibrücken. From an engraving it can be seen that the top of the platen is ridged as in the *Albion*, and that a projecting tongue of iron catches in the end of a pressing beam, when a lever underneath the apparatus is pressed. However, this is not a foot treadle but a hand lever, and it looks as if the press is a bench model. Knowing

Dingler's propensity for copying and altering British presses it may be that this type of small press, with foot operation, was common at one time.

Certainly the idea was not new, as the notion of combining a simple hinged press with a treadle had occurred to one 'M. N.' as early as February 1825, for he wrote to the *Glasgow Mechanic's Magazine* of that date with his suggestions for a 'new portable printing press'. This consisted of a table, a separate hinged board to carry types and fulcrums and levers. What is particularly interesting is that the illustration shows a double treadle (not described). The foot was to be placed on the first treadle to lift the bar to raise the levers and allow the type board to slide beneath the platen. The foot was then to be removed and applied to the other treadle 'which will produce a strong pressure upon the types'.

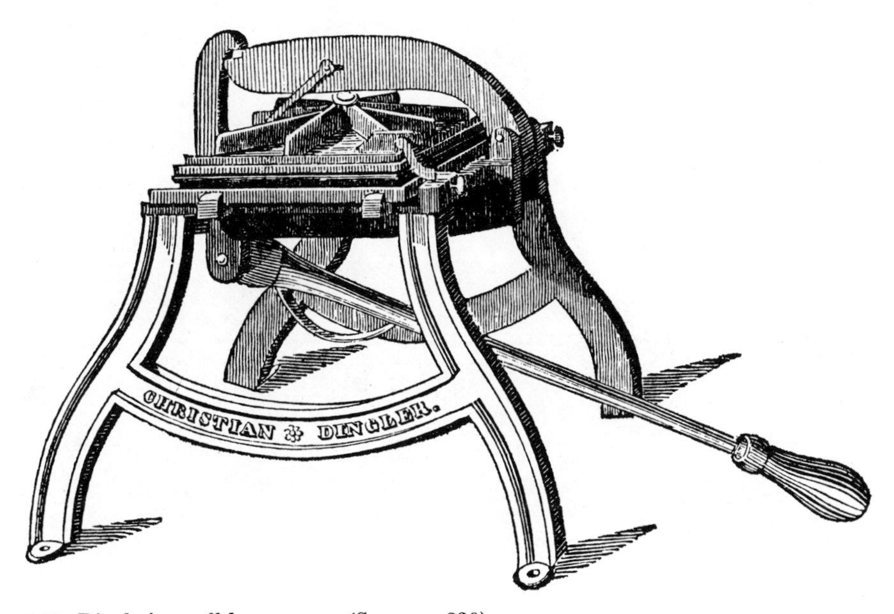

103. Dingler's small beam press. (See page 236)

A magazine which took a great interest in amateur presses was *The Bazaar*, published from 32 Wellington Street, the Strand, London. Comments by its contributor, P. E. Raynor, on certain amateur presses, were not appreciated by the manufacturers, and when in 1876 his assessment of presses was published in a booklet entitled *Printing for Amateurs*, he added: 'We can only say that in each case we have spoken as we have found in our own experience.'

Raynor provided helpful notes on the choice of a press, and, although he may not have realized it, one experience must have echoed that of the very earliest experimenters. As outlined in Chapter 1, screw presses existed in Gutenberg's time, but they were not necessarily appropriate for printing from type. Raynor warned against presses of a 'decidedly inferior make'—presses which obtained the impression by means of a screw, like an ordinary copying or bookbinder's press.

He had found that from their very nature they could not produce a presentable impression—'a screw from above descends on the platten in the middle, producing at best a very uneven impression and rendering expedition impracticable'.

Raynor then considered various presses, some of them comparatively well publicized, such as Berri's *People's* press, Jabez Francis's press, which he calls the *Everybody's*, and the Holtzappfel–Cowper *Parlour* press; but the booklet is particularly valuable as a source of reference to manufacturers whose names have long since been forgotten, and to entrepreneurs who, despite the disapproval of the printing trade, were willing to set the amateur up with a complete, if small, printing shop.

104. *Printing at home*, 1870

Very simple mechanisms were still being made in 1876 by such firms as Fairbairn & Co., of Smithfield, London. Their press was made of wood and consisted of two blocks with a wooden lever handle. The price was therefore low, and to print an area of 6 × 4 inches the amateur could get a press for 9*s*., 10 × 8 inches 18*s*. and for 12 × 10 inches £1 4*s*. Fisher's *Model* printing 'machine' (W. Fisher, 65 Broad-mead, Bristol) differed slightly in construction from the simple lever press. The bed, composed of a solid block of wood, slid in a groove beneath the centre, where pressure was applied by another block attached to a lever handle.

Clement Malins, 26 Marshall Street, Birmingham, and Ward's of 9 Goodham Hill, Burnley, both produced a press called the *Excelsior* (a favourite name), but they were distinct in structure and principle. Ward's press was on the ordinary hand-lever principle whereas Malins's worked as follows: the type was fixed in a box, paper placed on the platen and lowered on to the type, the whole being slid under an eccentric lever, which was turned over on to it, giving the impression. *The Bazaar* did not think much of either of these presses.

There was not much originality among manufacturers. A. N. Myers & Co., 12 Berners Street, London, sold two classes of press, the first being almost identical with Malins's, and the second the same as Berri's (which was a copy of the American army field press and is referred to later). Mr. Raynor's feelings about the simple wooden lever presses, which were not very far removed from the 'bellows press', are summed up in his report on Millikin and Lawley's presses, which were on the same principle as those of Myers: 'These presses are really useless for practical purposes, and may be set down in the category of toys.'

William Wightman, of Dewsbury Road, Leeds, apparently not worried by professional disapproval, advertised his amateur presses in the trade journal, the *Printers' Register*, during 1871. No description or name was given but the presses must have been fairly simple affairs if the price was anything to go by. A press to print 6×8 inches cost 12*s*., 9×7 inches—15*s*. and 10×8—18*s*. *The Bazaar* booklet throws a little more light on Mr. Wightman's products. The *Little Stranger* was a small iron press on the lines of the *Parlour* press, and the one to print 9×7 inches was called, perhaps appropriately, *The Novice.*

The hand stamp was still being classified as a 'press'—for example, *Baum's* 'printing press', which Jacques Baum & Co., of Birmingham, sold for a shilling. It was dismissed by Raynor—'This is a mere toy for printing a single line of type. It is a narrow trough of brass in which the type is fixed.' Heal so issued a warning about J. Theobald & Co., 20 Church Street, Kensington, who advertised a printing apparatus. 'Let no one be deceived', he wrote, 'by the grandiloquence of the announcement, "Complete Printing Apparatus", into believing that he can get a printing plant for half a crown.' The apparatus consisted of a small block of wood in which a single line of long primer type could be fixed and could be used as a hand stamp.

We see in these efforts the beginnings of the later ubiquitous 'John Bull' type of printing outfit—a box containing sets of rubber type, a wooden hand stamp with grooves into which to slide the type in the required order and an inking pad.

But much more advanced 'complete printing offices' were available, although Raynor thought this was an expensive way of getting materials. Such 'offices' were supplied by Francis and Wightman. Charles Morton, 'Type founder &c', of 167 City Road, London, advertised 'the Cosmopolitan Amateur Printing Office', No. 1 costing £5 10*s*., the basic item being a small *Albion* press at £2 10*s*., capable of printing an area of 8×5 inches. It was Raynor's opinion that at that size a small wooden press was cheaper and quite as effective. Morton's No. 2 'office', at £10, included an *Albion* press, at £4 10*s*., which printed double the size.

Competition must have been severe. T. W. Martin, of 89 Shoe Lane, London, supplied an amateur printing office for £6, including 'a real Model of the *Albion* Press used in large offices'. The press was priced at £3, but no details of size are given. The rest of these 'offices' consisted of founts of type, cases, chases and composing equipment, but Martin threw in an iron ink table. An odd aspect of Morton's small *Albion* presses was that they had an *Atlas* action.

The *Parlour* type of press, although ancient in origin, seems to have been peculiarly English in development, but at least one hinged table press was made in France, by Berthier et Cie, of Paris, who advertised it as *La Mignonette* in 1876.

From an engraving it looks a fairly simple type of mechanism. The smallest size sold for 30 francs.

For some reason the *Parlour* press did not catch on in America, where the small job platen became the most popular press for amateurs, although at first the iron hand press was used as the basis for small card presses. An acorn-shaped version was made by Samuel Orcutt, of Boston, which the Germans called the 'Yankee Karten Presse'.

Small presses were required by others besides professional printers or those wanting a hobby—the armed forces, for example. The Cincinnati Type Foundry produced its *Army* press, originally designed by Henry Barth as a portable newspaper press for Union troops in the field during the war between the States. It was also used for printing dispatches and orders, as was the similar Adams *Cottage* press. A British example, probably copied from the Americans, was the *People's Printing Press*, sold by D. G. Berri, of High Holborn. The press of this type consisted of a flat bed, to which a platen was hinged, and when turned down the whole was moved by a gear wheel under a fixed cylinder of which the pressure could be adjusted by screws. Berri claimed that his press had been used by the British Navy.

In the drive westward in the United States there was no inhibition about using any particular kind of press to print a newspaper in a new township and the Army or Cottage press was pressed into service as, for example, by the *Montesano Vidette* in Washington State in 1883.

One amateur press of a novel construction was the *Lowe* press, patented by Samuel W. Lowe, of Philadelphia, in 1856. It was based on a heavy conical roller (at first made of wood but later of iron) on a shaft rotating about a vertical spindle, somewhat reminiscent of the cylinders on Koenig's projected multiple machine. The type was placed on a cast-iron bed, at the end of which was an iron post which served as a fulcrum to a lever carrying the conical roller, which was able to move freely from side to side. As the roller was brought down to the bed of the press it pushed the paper on the tympan on to the type. By grasping the handle at the end of the lever the operator was able to rotate the roller over the top of the tympan and then return it to its original position after the impression had been made. By means of a spring the tympan raised itself and the printed paper was then taken off.

The *Lowe* press was first exhibited at the Fair of the American Institute held in 1857 at the Crystal Palace, New York. It was awarded a silver medal, the highest prize for a printing press, 'it being the cheapest, simplest and best press exhibited'. Sold in five sizes—5×6 inches at $5, 7×12 inches at $10, 12×14 inches at $16, 13×17 inches at $25 and 19×23 inches at $50, the press, it was claimed, could be easily managed by a child of eight and that 'one person alone can print from three to five hundred sheets an hour, and with the assistance of a boy to ink the types, seven hundred can be printed with ease.'

Despite all these encomia the *Lowe* press did not make headway. A few models survive and are used by enthusiasts, but it was the small version of the jobbing platen which made the greatest appeal to American amateur printers. It was only natural that somebody should think of scaling down Gordon's invention in the country of its origin and it is likely that the pioneer was B. O. Woods, who invented the *Novelty* press in about 1869, and had it marketed by Kelly, Howell and Ludwig of Philadelphia. A hand lever was used to operate the smallest sizes but a treadle attachment was provided for the bigger models, a characteristic which was to last until recent times.

105. 'Card and billhead' press made by Boston & Fairhaven Iron Works, 1871

Indeed, it is sometimes difficult to decide at which market—amateur or professional—some manufacturers were aiming. Perhaps they had none in mind but simply wanted to sell as many presses as possible, but the man who put the small jobber on the map, William Kelsey (born 1851), of Connecticut, was fairly sure of his target—the amateur, and preferably the young amateur. He later produced larger presses, and his successors continued to do so, but he became famous as the man who made printing presses for boys.

Kelsey first advertised in 1872 before it was even known whether his newly built press would work. As it failed to do so there was naturally considerable embarrassment. In respect of premature advertising he was to be followed half a century later by a British amateur press maker—D. A. Aspinall ('Mr. Adana').

But it is a matter of history that Kelsey eventually made his press work. He named it the *Excelsior*, presumably ignorant of the English press of that name, which, in any case, was based on different principles. He gradually developed the *Excelsior* until, in 1875, a toggle action and automatic inking system of the rotating disc kind was incorporated. The press was substantially the same as that which is still being sold by the Kelsey company of Meriden, Connecticut. The 5×8 and 6×10 sizes are available with either a handle at the front or a side-lever handle, but the others only with a handle at the front. 'Handle' and 'lever' are used without too much precision, but often the lever is a simple piece of metal, sometimes known as a stick lever, occasionally improved with a bulbous top, and sometimes it has a protruding horizontal handle. 'Handle' can also refer to one which is stirrup-shaped; others have curved ends.

Some confusion of names must have occurred as William Braidwood, of New York, made a card and job press, also called the *Excelsior*, and yet another press of the same name, sold by the Excelsior Printers Supply Company, in England, had no connection with Kelsey. Another company which did act for Kelsey then brought out its own *Model* press, which has been supplied by a number of firms since, including Squintani & Co., which advertised it in the Caxton Celebration Exhibition catalogue of 1877. The press continued to be made until the early 1960s.

Both in the United States and Britain entrepreneurs copied Kelsey. Joseph Watson, of New York, in 1875 called one press the *Centennial* and another *Young America*. J. Cook made the *Enterprise* and the *Victor*; Curtis and Mitchell the *Caxton* and the *Columbian* until the 1890s. The Golding firm, famous for its jobber, made the *Official* table press, which was mounted on a walnut base board (Plate LII). It ranged in size from the tiny *Junior*, with a $2\frac{1}{2} \times 3\frac{1}{2}$-inch chase (Plate LIII), to one with a 6×9-inch chase, which was characterized by elaborate decoration of the metal parts and a stirrup-shaped handle. The *Official* was also made in large sizes and could be driven by treadle or power.

The number of firms imitating Kelsey in the United States has been estimated at twenty-five, and the period in which they were most active was between 1880 and 1910. Their number declined as Kelsey absorbed his rivals, including the pioneer, B. O. Woods; J. Cook & Co.; Joseph Watson, and Curtis and Mitchell. Today there are only two other American firms making small presses. One is the well-known Chandler and Price, of Cleveland, Ohio, which is mainly concerned with full-sized equipment, but which does manufacture the *Pilot* platen press mainly for schools, distributing through agents. The *Pilot* dates back to 1886, but was redesigned in the 1950s. The name *Pilot* has been used by other manufacturers. The other firm is the Craftman Tool Company, of Boston, Massachusetts, which is responsible for the *Imperial* (8×5), the *Superior* ($10 \times 6\frac{1}{2}$) and the *Monarch* (12×9).

The Sigwalt Company, of Chicago, made hand-lever presses from the early 1900s until about 1962, ranging in size from $4 \times 2\frac{1}{2}$ inches to 9×6 inches, and patterned closely on the Golding *Official* presses.

Much of the research on the smaller American presses for boys has been carried out by Martin Speckter, of New York, who has an unrivalled collection of printing presses, including a wide range of miniatures. These include examples of the *Daisy*, which prints formes 2×3 inches, and which was sold for a dollar by Sears and Roebuck in 1900; the so-called *Giant* (patented 1876), with a $2 \times 2\frac{3}{4}$-inch chase (Plate LIV); and a toy press of stamped tin, made by the former Elm City Toy Company, of Hartford, Connecticut. Although supposed to be a toy, this press is a typical self-inking clamshell platen with a revolving ink disc. Other tiny presses, including the *Giant*, do not incorporate an ink disc. The *Giant* has a curved ink plate, and the *Daisy* and the *Baltimorean* small-size press are inked by a hand roller, which rests on an inking plate above the forme when not in use.

In England, C. Morton of City Road, London, who it has been noted produced a small *Albion*, brought out a rival to the *Model*, and the Birmingham Machinists Company made a version of the German *Simplicissimus* bench model press. This was rather odd-looking but worked quite well. Thomas Taylor & Sons, of Leicester, make the *Favourite*, with a 9×6-inch platen, and the Adana Company are responsible for two separate models of the small jobbing platen today.

The West German Höhner Maschinenfabrik make small presses which bear various names, according to the country in which they are distributed, and in sizes of $10 \times 6\frac{1}{2}$ and 12×9. *Pilot* is the name given by the United States distributor to the $10 \times 6\frac{1}{2}$, and other names such as *Boston* are used except in the United States. These presses are close copies of the Golding presses.

The *Parlour* press never died out in Britain, and as late as 1919 Adams Brothers, of Daventry, were still making a very simple wooden press for boys. Basically it consisted of a bed to which was hinged a platen at one end and a lever at the other. There were no refinements. Type was locked in the bed and inked with a hand roller, paper was held on the platen by whatever means the operator could devise, the platen was lowered on the type and the lever brought down on top of the platen to exert pressure.

In 1922 Donald A. Aspinall, of Twickenham, who had been working in a firm selling racing tips by mail, found himself out of a job, and conceived the idea of making model printing presses. It is a matter of speculation as to why he chose the *Parlour* press type rather than, say, the jobbing platen, but there is reason to believe he had seen a reference to the *Parlour* press in a mechanics' journal, and was impressed by its simplicity. In any case he did not have the resources to make anything more complicated.

At his lodgings he made a little press, mainly from wood, but with a metal attachment forged by a local blacksmith. His landlord, an obviously interested party, suggested that he advertise it in the *Model Engineer* at 45*s*. or 50*s*., including some type. The response to the advertisement was embarrassing—buyers sent money to a man who had made a prototype only. He was so worried that he reported the matter to the police, and asked what he should do. The reply was blunt: 'Make the presses'. Aspinall therefore rented a stable loft in Twickenham and got to work to fulfil the orders.

The prototype had been a version of the *Parlour* press which could fit into a cigar box, and it had to be inked by hand. The production model was still small, with an inside chase of $4\frac{3}{4}\times7$ inches, and consisted simply of a platen hinged to a wooden box. On top of the platen was a wooden lever with a handle at one end and tapered towards the other. When the platen was closed over the top of the box the tapered end of the lever caught under a metal arch screwed to the far end of the box, and in this way exerted pressure. However, there was one technical advance—an attempt to achieve automatic inking.

Protruding from the end of the box was a metal plate on which ink could be spread, and at each side of the platen was a curved bar, between the ends of which was a composition roller. The effect of closing the platen was to push the roller over the ink plate; opening the platen to roll it over the type in the box (Plate LVI).

By 1928 the press had not changed basically, although it was better made, and available in sizes of 9×5 inches and $11\times8\frac{1}{2}$; but during that year the new 'all steel' model was developed. The inside chase was still $4\frac{3}{8}\times7$, and the price still 45*s*., and although there was no advance in technique, the fittings were more refined, and movable grippers had been added to keep the paper in position.

At the same time it was decided to reach out to the very juvenile market and make a tiny pocket press. Advertised at 5*s*. 6*d*., it was later sold for 10*s*. 6*d*. The press, which had an inside chase of $3\frac{5}{8}\times2\frac{3}{8}$ inches, was of extreme simplicity, not much more than a 'bellows' press, but with automatic inking. The whole affair fitted on a wooden base (Plate LVII). This small press was improved by about 1931, being made throughout of pressed steel.

A quarto-sized press was produced in 1934, made of steel and iron, and held on a wooden base. An attempt had been made to overcome the problem of uneven pressure by an improved lever which covered a bracket extending over a wide area of the platen. This press was sold at £2 18*s*. 6*d*.

It is to be expected that Adana's success would stimulate competition, but the only known rival to the Adana 'flatbed' which can be traced is the *Ajax*, made by the Ajax Engineering Company, of Southend-on-Sea. This had a platen impression adjustment consisting of springs and wing nuts, by which the platen could be adjusted either to or from the type bed. There was also only one roller arm so that larger sheets of paper could be printed.

The development of the *Adana* flatbed continued, and by 1937 for £4 10*s*. the amateur printer could buy a press with a $7\frac{1}{4}\times9\frac{3}{4}$-inch chase, which resembles the press of today. By a stroke of inspiration the revolving ink disc from the jobbing platen was incorporated, and this could be bought separately for £1 7*s*. 6*d*. A single arm roller enabled any length of paper to be printed.

Today's press is known as the *Adana* horizontal platen, and is made completely of metal with a chase of $9\frac{3}{4}\times7$ inches. While basically the same as the first of its kind, it is more elaborate, efforts having been made to make it more efficient. There are two composition inking rollers, an ink disc, a special pressure adjustment on the lever, an adjustable platen and sliding gripping fingers. It is popular with typographic designers who can use it for experimental work. During the 1939–45 war Aspinall produced a special model, without ink disc, and designed to be collapsible, for dropping to underground groups behind enemy lines.

Aspinall had become aware of the attractions of the jobbing platen type of press, and by 1933 began his first attempt in this direction. His first cast-iron model had an inside chase of $3\frac{5}{16} \times 1\frac{13}{16}$ inches. There were no gripper fingers, and the press sold for 28*s.* 6*d.* (Plate LIX). By 1936 the price was 38*s.* 6*d.* and in about 1936 a bigger model, with a $5\frac{3}{4} \times 3\frac{1}{8}$-inch chase, was produced for £4 12*s.* 6*d.* This had extras such as a rider roller, treadle and ink duct. In 1938 an attempt was made to sell a press at the very low price of 18*s.* 6*d.* (Plate LVIII). It was made of sheet metal with a wooden handle and gripper fingers. Chase size was $3\frac{13}{16} \times 1\frac{5}{16}$ inches. The press turned out to be a completely uneconomic proposition and the partially cast-iron No. 2 High-speed, with chase size of $5\frac{3}{4} \times 3\frac{1}{8}$ inches, at £4 12*s.* 6*d.*, was introduced. In 1939 the cast-iron $10 \times 7\frac{1}{2}$-inch model, at £7 17*s.* 6*d.*, made its appearance. Today's models have chases of 5×3 and 7×5 inches, and are consequently known as such.

During the war Aspinall gradually retired as the result of ill health, and the firm was acquired by F. P. Ayers, an engineer who had first been associated with Adana in 1924, when he was asked to do some machinist's work on gears. He subsequently designed a number of the presses, including the T/P power press.

The name *Adana* is explained by the fact that Aspinall had served in the army during the First World War for part of the time at Adana in Turkey (an ancient town often disputed during the Crusades), and fancying a similarity between his own initials and the name of the town decided to call his presses after it. In publicity material he often called himself 'Mr. Adana'.

A rival to the *Adana* jobbing platen was the *Lilliput* 'hand printing machine', made by Acme Printing Supplies Company, but this made very little impact on the amateur or professional field.

The lack of precision in naming various presses is no more apparent than when discussing 'card', 'billhead' and 'jobbing' presses. A small press, of curious design, with a platen of only $7 \times 11\frac{1}{2}$ inches, had been patented by A. & B. Newbury, of Coxsachie, New York, in 1859, and was called the *Country Jobber*. It was a mixture of the hinged press and jobbing platen. The inking disc hung from the end of the horizontal bed, and as the lever was pulled down to close the platen an inking roller crossed the type and moved downward over the disc. A. N. Kellogg, of Washington Street, Chicago, produced a 'new style improved' version of the *Country Jobber* in 1863, renaming it the *Newbury* blank and card press. An advertisement shows a boy operating the press, hand feeding from a table protruding on a bent iron shaft.

In reality, all these presses were versions of the hinged press, some being horizontal and others vertical, and some more advanced than others with automatic inking and treadle operation. This becomes clear if the 'card and billhead' press made by the Boston and Fairhaven Iron Works, Fairhaven, Massachusetts, in 1871, is studied. It is a fairly skeletal structure, but has a simple ink disc and foot treadle. A very similar press produced in Paris by Berthier in 1876, called *L'Abeille*, has no ink disc. This must have been for reasons of economy as Berthier could not have been unaware of the inking disc as he was the agent for the *Minerva* jobbing platen.

To some extent, therefore, any discussion of small presses covers the same ground as that on the jobbing platen, but there was a specialist demand for a press which could literally churn out small printed cards or tickets by turning a handle. George Gordon, the pioneer jobbing platen maker, considered this point as early as 1853, and brought out a bench card press called the *Lightning*, which was turned by a crank handle and was claimed to be able to print cards from a continuous roll of stock at the rate of 8,000 to 10,000 an hour. Built-in shears cut the card from the roll after each impression.

Gordon was followed by Franklin L. Bailey, who, after patenting a card press in 1857, assigned the patent to R. Hoe & Co. The fly-wheel and ink cylinder rotated on the same shaft, which was turned by a crank handle. The platen was a mere 4×5 inches, and a speed of 1,000 to 2,000 cards an hour was claimed. Feeding of separate cards was automatic, and the finished products dropped into a trough. In 1865 the press was adapted for ticket printing, and was known as the Hoe patent numbering card press. An additional mechanism for numbering the tickets consecutively and depositing them in a tray in numerical order was added. It was capable of numbering up to 10,000.

106. A. N. Kellogg's 'new style improved' Country Jobber, 1863. (See page 245)

An example of the *Official* card press, with a vertical impression, made by the Feuerstein Company, of Chicago, *circa* 1880, survives in the Speckter collection. The type was locked in a small chase held upside down. Every revolution of the crank handle brought the type down on to the paper stock. Inking was by disc.

In France the pioneer of this type of press had been G. Leboyer and among others, Berthier, Poirier and Pierron et Dehaitre were responsible for similar machines. A. Magnand, or his successor, J. Hariel, of Paris, constructed an elaborate 'cartes de visite' machine, an example of which survives in British ownership.

While the uninitiated may think that printing presses are large affairs for printing newspapers and books, there have always been those who required presses for small, and specialist purposes, and, apparently, there has nearly always been somebody willing to make them. No doubt the full-size travelling press built for Peter the Great in 1723, which is in the State Historical Museum, Moscow; the press for 'great lords', advertised by the Breitkopf foundry, which occupied a box 3 feet high by 2 feet wide; and even 'one of the smallest presses in existence' owned by the 'old Duke of Norfolk', mentioned by John Johnson in his *Typographia* (1824), were conventional screw and bar presses, made by joiners and smiths, but the nineteenth century offered new opportunities for the ingenious mechanic to break away from the old techniques. It was when the transformation of society itself, particularly in the United States and Britain, began that special presses were required to produce the multifarious small pieces of printing used in modern life, and to reproduce in quantity (even though small) the creative work of an increasingly articulate democracy.

The printing antiquary, or even sentimentalist, who is also an amateur printer, often wishes to use a distinctive, historical press. This accounts for the demand for small *Albions* and other well-known iron hand presses. The amateur printer, who is primarily concerned with results, will, on the other hand, be most attracted to a press which is most suited to his work. Thus he may feel that a self-inking jobbing platen is best or, increasingly, he may be seen turning to a modern, automatic, precision flatbed proofing press, which derives in the long run from Faustus Verantius, the *Ulverstonian*, the *Belper*, the *Army* and the *Cottage* presses.

107. Hoe's version of Bailey's card press. (See page 246)

108. Soldan & Co's proofing press, *c.* 1900

APPENDIX II

THE PROOF PRESS

One kind of proof is a trial impression from composed type and blocks taken for checking and correction and, according to Stower, an empty or unemployed press was kept in every printing office for pulling proofs. The emergence of the iron hand press made little difference to this procedure, but when the cylinder press came into use various types of proof press were specially developed. Strictly speaking, any press may be used as a proof press, but owing to the methods of making up pages of type for letterpress printing some presses are more convenient than others.

Consideration should also be given to the printers' galley, a flat oblong tray to hold composed type matter before it is divided up into page form for books or columns for newspapers. A 'galley' proof is a rough proof taken of such type on a slip of paper about twenty inches long. This does not have to be particularly well printed as long as the result is readable, and a proof could be made by simply pressing a piece of the inked type by hand—although this would not be very satisfactory.

In the earliest days a galley could also be a square box with, sometimes, a sliding board to assist in the transfer of type from the galley to the imposing stone. If type were taken direct to the wooden press on this board for proofing, it would have meant adjustment of the platen because of the extra depth of the printing element.

Specialized newspaper printing produced the long galley for column proofing, and a surviving wooden press in the Hunterian Library, University of Glasgow, was formerly fitted with an oblong mahogany plank screwed to the bottom of the platen for the proofing of long, narrow printing surfaces. Altering wooden presses for purposes for which they were really not suited must have been a tedious business and convenience would have soon dictated the development of a separate proofing device.

Compositors thus made a simple press not far removed from that suggested by Faustus Verantius in the seventeenth century (see Chapter 7). Typeset matter was placed between two type-high parallel bars. After inking, a piece of paper and a blanket were placed on top of the type and an iron cylinder was run along the bars over the type to produce a proof. A press on these lines is referred to in M. D. Fertel's *La Science pratique de l'imprimerie* in 1723. Small pieces of printing called 'factums' were proofed by a 'rouleau'.

The credit for producing a commercial version of this simple device is usually given to Stephen Tucker, of R. Hoe & Co., and from 1844 onwards the Hoe firm made presses of this nature for the printing trade. One press made for Dr. Miles, of Elkhart, Indiana, who made patent medicines, was given to country newspapers in exchange for advertising space for Dr. Miles's remedies. The press has 'Miles' Nervine' cast in raised letters at one end and 'Miles' Heart Cure' at the other. Another press of this kind was the Whipple *Economic* press, for which a felt blanket for the cylinder could be bought for an extra $1·50. Eventually the idea of a surface for the roller of felt or indiarubber took hold, saving the time taken in placing a blanket on top of the proof paper.

In England, Francis Donnison & Son, Northumbrian Steam Printing Works, Newcastle upon Tyne, advertised in 1870 a cheap and strong galley press, with stand, cylinder, ink roller and slab for £6 6*s*.

As hand presses were used for proofing it followed that manufacturers would make special versions for galley proofs and, as described in Chapter 4, Wood and Sharwoods produced a simple lever press called the *Columbian* for this purpose. A similar, nameless, press was produced in the 1850s by Harrild's, and Frederick Ullmer advertised his 'improved *Columbian* galley press' some forty years later.

But proofs of whole pages were also wanted—by publishers, and by printers requiring to know that all was well before locking several page formes in a chase. The hand press suited this requirement, but several firms made special page-proof portable presses, such as that issued by the Boston Type Foundry in 1876. The cylinder was placed between two wheels so that it could be run over an inked forme. Pressure of the cylinder against the forme was performed by a handle attached to two upright posts containing springs.

The idea of taking the press to the type instead of the type to the bed of the press, as exemplified by the Boston Type Foundry device, persisted. In 1953 the *Ashton* portable proofing press, invented by Sidney Ashton, of Manchester, brought together in one portable unit a system of ink rollers, a roll of paper and an impression roller. The press is rolled once over the galley for inking and a second time for the printing impression.

Apart from presses which could be used for producing proofs for simple checking there were those which were needed to produce superior proofs for engravers, either for submitting to publishers or for use in making ready on the printing machine. This requirement is one reason why the solid iron hand press not only survived but continued to be made well into the twentieth century. The advent of process engraving (or the photo-mechanical technique of making relief blocks) made accurate proof presses all the more necessary, and led press manufacturers to stress the strength of their hand presses. T. Matthews, noted as a manufacturer of *Albion* presses, stated in 1896: 'Our presses, being extra strong and powerful, are used by most of the leading firms.'

There were also such extra-heavy presses as the *Herculean* and the *Lion*, but the manufacturer who really carried the heavy-duty hand press to its fullest extent was Paul Shniedewend, of Chicago, whose British agents were Penrose & Co., suppliers of process-engraving equipment. In 1898 he was advertising his *Reliance* (a *Washington*-type press) as 'a new press of unequalled strength and rigidity for proving half-tone and process cuts'. The 'New A' model weighed 975 lb. and the 'New B' 1,640 lb., but by the turn of the century he had produced the *Lion* (presumably not knowing of the British-built *Lion* press). This was no less than 3,000 lb. in weight. It was surpassed by the *Mammoth* (weight 4,200 lb.), 'a proving press

Price, beautifully finished and fitted, £1 11s. 6d.

WITH the use of this simple Machine proofs may be pulled as well and as quickly as with the most expensive galley-press. It possesses the additional advantage of being capable of taking proofs of pages lying on the imposing surface. Where a galley is used, the spring on the roller compensates for the thickness of the bottom o the galley. The space between the wheels is 12 inches.

Price, in Box Complete .. .. .. £2 2s.

109. J. M. Powell's advertisement for a hand proofing press

of enormous power', and finally, in 1902, by the *Mastodon* (weight 5,000 lb.), 'the largest and most powerful proving press ever built'.

At the other end of the scale Shniedewend produced the *Midget*, with a platen of $14\frac{1}{2}\times18\frac{1}{2}$ inches, and the *Baby* (platen 10×8 inches), for 'wood engravers, small firms or amateurs'. An interesting aspect of the *Baby* is that it has no rounce mechanism, the bed being pulled in and out by hand, in the manner of some early presses.

Shniedewend died in 1911 and his firm was taken over successively by the Williams-Lloyd Machine Co. of Chicago and New York and the United Printing Machine Co. of Boston. Shniedewend's presses were widely used in America and Europe and were flattered by imitations, including Hewitt's *Acme* and the Hunter *Ideal*.

However, other manufacturers stressed the value of the cylinder type of press. Soldan & Co., in 1902, for example, advertised the *New Age* proofing press. No half-tone printer should be without one, they urged. It gave a proof equal to the finished job without any making ready being necessary, and although they may not have realized the significance at the time, they added: 'The heaviest type formes and finest half-tone work can be printed on this machine quite apart from proof pulling.' In a sense, therefore, the ordinary printing press had been transformed into a proofing press, and, as it developed, the proofing press, because of its efficiency, could be used for printing.

In 1909 Robert Vandercook, of Chicago, whose name, to some extent, was to become synonymous with proof presses, produced his first press intended for small job formes. This was of unusual structure, being a rocker type with a half-cylinder with teeth at each end designed to mesh with other teeth in rails on the bed, so that the cylinder could not slip. The rocking motion was provided by a handle. Vandercook went on to produce a more conventional galley proof press in 1911, based on the large roller, with a separate inking device; and then settled down from 1914 onwards to a travelling carriage type of press, which became increasingly larger and more diverse, eventually being power-driven.

Many firms built proof presses from the turn of the century onwards, and with the development of photographic techniques in the 1920s and 1930s a new kind of proof was required. This was the reproduction proof pulled from a letterpress forme for copying by camera and transfer to a printing plate. The need for high-quality production led to even further precision on proof presses, and the transfer of the actual printing to pressmen who were more familiar with inking, impression and make-ready.

The proof press, in one shape or another, will be required while metal type is used for composition. As photographic techniques develop further it is possible that the relief proof press will go into a decline, but will find a home among those requiring precision printing for short runs.

BIBLIOGRAPHY

American Dictionary of Printing and Bookmaking. New York, 1894.

Annison-Duperron, E. A. J. *Description d'une Nouvelle Presse*. Paris, 1783.

Benedikz, Benedikt S. *The Spread of Printing in Iceland*. Amsterdam, 1969.

Berry, W. Turner. *Augustus Applegath: Some Notes & References*. London. Journal of the Printing Historical Society, No. 2, 1966.

—— *The Autobiography of a Wooden Press*. London. *Typographica*, No. 8.

Bigmore, E. C., and Wyman, C. W. H. *A Bibliography of Printing*. London, 1880.

Bloy, Colin. *A History of Printing Ink, Balls and Rollers 1440–1850*. London, 1967.

Burch, R. M. *Colour Printing & Colour Printers*. London, 1910.

Burke, Jackson. *Prelum to Albion: A History of the Development of the Hand-press from Gutenberg to Morris*. San Francisco, 1940.

Chambers, David. *An Improved Printing Press by Philippe-Denis Pierres*. London. Journal of the Printing Historical Society, No. 3, 1967.

Dieterichs, Karl. *Die Buchdruckpresse von Johannes Gutenberg bis Friedrich Koenig*. Mainz, 1930.

Easton, John. *Postage Stamps in the Making*. London, 1949.

Eastwood, A. *A Short Treatise on Stereotyping*. Burnley, 1901.

Eckman, James. *The Heritage of the Printer*. Philadelphia, 1965.

Fertel, Martin Dominique. *La Science pratique de l'imprimerie*. Saint Omer, 1723.

Frey, A. *Manuel français de typographie*. Paris, 1835.

Gaskell, P. *A Census of Wooden Presses*. London. Journal of the Printing Historical Society, No. 6, 1970.

Gibson, Peter. *Modern Trends in Letterpress Printing*. London, 1966.

Green, R. *The Iron Hand press in America*. Rowayton, Conn., 1948.

—— *A History of the Platen Jobber*. Chicago, 1953.

Haas, Wilhelm. *Beschreibung einer neuen Buchdruckerpresse, 1772*. Basel, 1790.

Hamilton, Milton W. *Adam Ramage and his Presses*. Portland, Maine, 1942.

Hansard, T. C. *Typographia*. London, 1825.

Hargreaves, Geoffrey D. 'Correcting in the Slip: The Development of Galley Proofs', *The Library*, December 1971, Vol. XXVI, No. 4.

Hart, Horace. *Charles, Earl Stanhope and the Oxford University Press*. Reprinted from Collectanea III, 1896, of the Oxford Historical Society, with notes by James Mosley. London. Printing Historical Society, 1966.

Hoe, R. *A Short History of the Printing Press*. New York, 1902.

Howe, Ellic. *Newspaper Printing in the Nineteenth Century*. London. Privately printed, 1943.

Hutchings, R. S. *Josiah Warren, Anarchist and Inventor*. London. *The Black Art*, Vol. 2, No. 3, Autumn 1963.

Isaacs, George. *The Story of the Newspaper Printing Press*. London, 1931.

Johnson, J. *Typographia*. London, 1824.

Kainen, Jacob. *George Clymer and the Columbian Press*. New York, 1950.

Karolevitz, Robert F. *Newspapering in the Old West*. Seattle, 1965.

Lilien, Otto M. *History of Industrial Gravure Printing up to 1900*. London, 1957.

The Ernest A. Lindner Collection of Antique Printing Machinery, The Weather Bird Press, Pasadena, 1971.

Liveing, Ed. *The House of Harrild 1801–1948*. London, 1949.

Luckombe, Philip. *A Concise History of the Origin and Progress of Printing*. London, 1770.

Macmillan, Fiona. *The Spread of Printing in New Zealand*. Amsterdam, 1969.

Madan, Falconer. *Early Representations of the Printing Press*. *Bibliographia*, 1895.

Mason, J. H. *A Selection from the Notebooks*. Leicester, 1961.

Monet, A.-L. *Les Machines et Appareils Typographiques*. Paris, 1878.

Moran, James. *A Brief Essay on the Printing Press*. Lintzford, co. Durham, 1963.

—— *A Condensed History of the Relief Printing Press*. London. *The Black Art*, Vol. 3, No. 4, 1964/5.

MORAN, JAMES. *Stephen Austin of Hertford*. Hertford, 1968.
—— *The Columbian Press*. London. Journal of the Printing Historical Society, No. 5, 1969.
—— *The Development of the Printing Press*. London. Journal of the Royal Society of Arts, April 1971. Reprinted by the Wynkyn de Worde Society, July 1971.
MORES, EDWARD ROWE. *A Dissertation on English Typographical Founders and Founderies*. London, 1778. New edition, edited by H. Carter and C. Ricks, Oxford, 1961.
MOSLEY, JAMES. *The Press in the Parlour*. London. *The Black Art*, Vol. 2, No. 1, 1963.
MOXON, JAMES. *Mechanick Exercises: or, the Doctrine of Handyworks. Applied to the Art of Printing*. London, 1683–4. New edition, edited by Herbert Davis and Harry Carter, London, 1958.
NIEPP, LUCIEN. *Les Machines à Imprimer depuis Gutenberg*. Paris, 1953.
OLDENOW, KURT. *The Spread of Printing in Greenland*. Amsterdam, 1969.
PAPILLON, J. M. B. *Traité historique et pratique de la gravure en bois*. Paris, 1766.
PIERRES, PHILIPPE-DENIS. *Description d'une nouvelle Presse d'Imprimerie*. Paris, 1786.
PLANTIN, CHRISTOPHER. *Dialogues françois* (1567), translated into English as *An Account of Calligraphy and Printing* and edited by Ray Nash. Cambridge, Mass., 1940, and Antwerp, 1964.
POLSCHER, A. A. *The Early Wooden Hand Printing Press in the United States*. London. *The Black Art*, Vol. 3, No. 4, 1964/5.
POTTINGER, DAVID T. *The History of the Printing Press* (in *A History of the Printed Book*). *The Dolphin*, New York, 1938.
Printing Apparatus for the Use of Amateurs. Holtzappfel & Co., 1846. Reprinted and edited by James Mosley and David Chambers. Pinner, Middlesex, Private Libraries Association, 1971.
RINGWALT, J. LUTHER. *American Encyclopaedia of Printing*. Philadelphia, 1871.
ROBLIN, FRED. *Printing Press Development, 1450–1965. The American Pressman*, November 1965, Vol. 75, No. 11, Tennessee.
SAVAGE, WILLIAM. *Practical Hints on Decorative Printing*. London, 1822.
—— *Dictionary of the Art of Printing*. London, 1841.
SCHOLEFIELD, G. H. *Newspapers in New Zealand*. Wellington, 1958.
SILVER, ROLLO. *The Costs of Mathew Carey's Printing Equipment*. Studies in Bibliography, Vol. 19, 1966. Charlottesville, Virginia.
—— *The American Printer 1787–1825*. Charlottesville, 1967.
—— *Efficiency Improved: the Genesis of the Web Press in America*. Proceedings of the American Antiquarian Society, October 1970. Worcester, Mass.
SMITH, CHARLES MANBY. *The Working Man's Way in the World*, 1857, reprinted with a preface and notes by Ellic Howe. London, Printing Historical Society, 1967.
SMITH, JOHN. *Printers' Grammar*. London, 1787.
SOUTHWARD, JOHN. *Progress in Printing and the Graphic Arts during the Victorian Era*. London, 1897.
SPECKTER, MARTIN. *Typoddities*. New York. *TYPEtalks*, No. 168, 1970.
STONE, REYNOLDS. *The Albion Press*. London. Journal of the Printing Historical Society, No. 2, 1966.
STOWER, CALEB. *The Printers' Grammar: or, Introduction to the Art of Printing*. London, 1808.
THOMAS, ISAIAH. *The History of Printing in America*. Worcester, Mass., 1810, and Albany, 1874.
TIMPERLEY, CHARLES H. *A Dictionary of Printers and Printing*. London, 1839.
VERANTIUS, FAUSTUS. *Machinae Novae*. Venice, 1616.
WATSON, JAMES. *History of the Art of Printing*. Edinburgh, 1713.
WENTZ, ROBY. *Eleven Western Presses*. Los Angeles, 1956.
WILSON, CHARLES, and READER, WILLIAM. *Men and Machines: a History of D. Napier and Sons Engineers Ltd. 1808–1958*. London, 1958.
WILSON, FREDERICK J. F. *Typographic Printing Machines and Machine Printing*. London, 1879.
—— and GREY, DOUGLAS. *A Practical Treatise upon Modern Printing Machinery and Letterpress Printing*. London, 1888.
WROTH, LAWRENCE C. *The Colonial Printer*. New York, 1931; Charlottesville, 1964.
ZONCA, V. *Nuovo teatro di machine et edificii per uarie et sieure operationi*. Padua, 1656.

GENERAL INDEX

INDEX OF PRESSES AND MACHINES

THE way in which printing presses and machines are designated presents a number of problems. Some have special trade-names and others are known after their inventor or manufacturer in a general way. Imitation led to confusion in nomenclature and some manufacturers produced a range of machines, within which there are specialized names for, say, particular sizes. Additionally, presses can be used for different purposes and thus duplication can arise when any attempt is made to index according to classification. The following index reflects these problems. To simplify matters all entries are in roman type. Where a press, such as the Stanhope, has been imitated but has not been given a separate name, it is recorded under the original. Where a special name has been given (Herculean Albion, for example) it is mentioned as such. Some names are duplicated—the 'Baby' could be used either as an amateur press or as a proof press.

HAND PRESSES

CYLINDER MACHINES

BED AND PLATEN PRESSES

JOBBING PLATENS

ROTARY MACHINES

TYPE-REVOLVING MACHINES

REEL-FED FLATBED MACHINES

SHEET-FED ROTARY MACHINES

MINIATURE, 'TOY', AMATEUR AND CARD PRESSES

PROOF PRESSES

Lithography: Braun-Brumfield, Inc.
Binding: Braun-Brumfield, Inc.
Paper: Warren's 1854, 60#

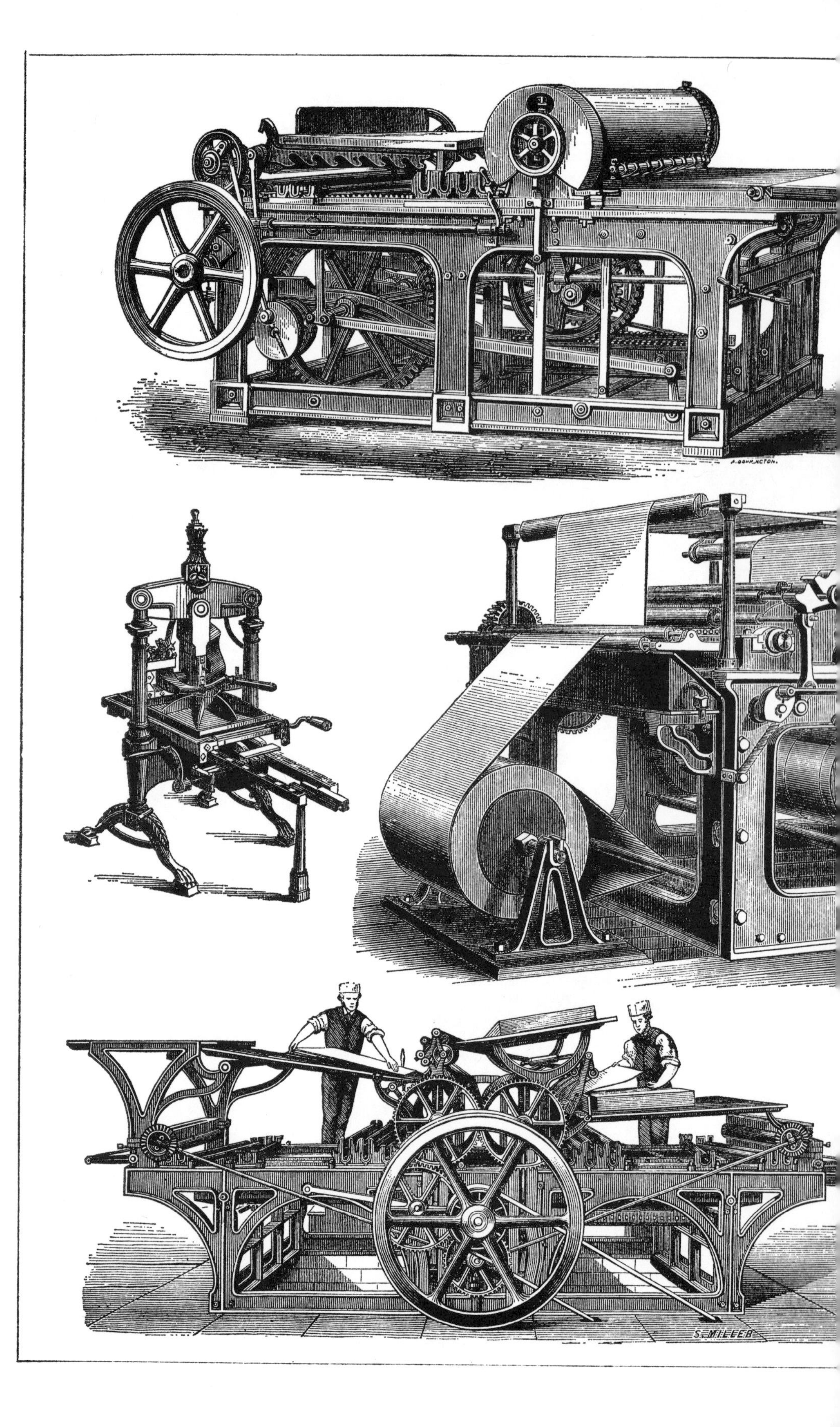
S. MILLER

MILLER & RICHARD